Assessing the Nation's Earthquakes

The Health and Future of Regional Seismograph Networks

National Research Council

ASSESSING THE NATION'S EARTHQUAKES
The Health and Future of Regional Seismograph Networks

Panel on Regional Networks
Committee on Seismology
Board on Earth Sciences and Resources
Commission on Geosciences, Environment, and Resources
National Research Council

NATIONAL ACADEMY PRESS
Washington, D.C. 1990

NOTICE: The project that is the subject of this report was approved by the Governing Board of the National Research Council, whose members are drawn from the councils of the National Academy of Sciences, the National Academy of Engineering, and the Institute of Medicine. The members of the committee responsible for the report were chosen for their special competences and with regard for appropriate balance.

This report has been reviewed by a group other than the authors according to procedures approved by a Report Review Committee consisting of members of the National Academy of Sciences, the National Academy of Engineering, and the Institute of Medicine.

The National Academy of Sciences is a private, nonprofit, self-perpetuating society of distinguished scholars engaged in scientific and engineering research, dedicated to the further-ance of science and technology and to their use for the general welfare. Upon the authority of the charter granted to it by the Congress in 1863, the Academy has a mandate that requires it to advise the federal government on scientific and technical matters. Dr. Frank Press is president of the National Academy of Sciences.

The National Academy of Engineering was established in 1964, under the charter of the National Academy of Sciences, as a parallel organization of outstanding engineers. It is autonomous in its administration and in the selection of its members, sharing with the National Academy of Sciences the responsibility for advising the federal government. The National Academy of Engineering also sponsors engineering programs aimed at meeting national needs, encourages education and research, and recognizes the superior achievements of engineers. Dr. Robert M. White is president of the National Academy of Engineering.

The Institute of Medicine was established in 1970 by the National Academy of Sciences to secure the services of eminent members of the appropriate professions in the examination of policy matters pertaining to the health of the public. The Institute acts under the responsibility given to the National Academy of Sciences by its congressional charter to be an adviser to the federal government and, upon its own initiative, to identify issues of medical care, research, and education. Dr. Samuel O. Thier is president of the Institute of Medicine.

The National Research Council was organized by the National Academy of Sciences in 1916 to associate the broad community of science and technology with the Academy's pur-poses of furthering knowledge and advising the federal government. Functioning in accordance with general policies determined by the Academy, the Council has become the principal oper-ating agency of both the National Academy of Sciences and the National Academy of Engi-neering in providing services to the government, the public, and the scientific and engineering communities. The Council is administered jointly by both Academies and the Institute of Medicine. Dr. Frank Press and Dr. Robert M. White are chairman and vice chairman, respectively, of the National Research Council.

Support for this project was provided under general funds for the Committee on Seismol-ogy through the following agencies: the National Science Foundation, the Department of Energy, the U.S. Geological Survey, the U.S. Nuclear Regulatory Commission, the Federal Emergency Management Agency, the National Aeronautics and Space Administration, and the Defense Advanced Research Projects Agency.

COVER: Art courtesy of Tanya George, Center for Earthquake Research and Information, Memphis State University. Depicted is the transmission of seismic data from national network stations to a satellite in stationary orbit. The data are then relayed to the National Earthquake Information Center in Golden, Colorado.

Library of Congress Catalog Card Number 90-61573
International Standard Book Number 0-309-04291-7

Additional copies of this report are available from

National Academy Press
2101 Constitution Avenue, N.W.
Washington, D.C. 20418

S157

Printed in the United States of America

PANEL ON REGIONAL NETWORKS

ARCH C. JOHNSTON, Memphis State University, *Chairman*
WALTER J. ARABASZ, University of Utah
GILBERT A. BOLLINGER, Virginia Polytechnic Institute and State University
JOHN R. FILSON, U.S. Geological Survey
ROBERT B. HERRMANN, St. Louis University
LUCILE JONES, U.S. Geological Survey
HIROO KANAMORI, California Institute of Technology

NRC Staff

WILLIAM E. BENSON, *Consultant*
BARBARA W. WRIGHT, *Senior Program Assistant*

Liaison Members

RALPH W. ALWINE III, Defense Advanced Research Projects Agency
CLIFFORD ASTILL, National Science Foundation
MIRIAM BALTUCK, National Aeronautics and Space Administration
MICHAEL A. CHINNERY, National Oceanic and Atmospheric Administration
JOHN G. HEACOCK, Office of Naval Research
LEONARD E. JOHNSON, National Science Foundation
GEORGE A. KOLSTAD, U.S. Department of Energy
PAUL F. KRUMPE, Agency for International Development
RICHARD A. MARTIN, JR., U.S. Bureau of Reclamation
UGO MORELLI, Federal Emergency Management Agency
ANDREW J. MURPHY, U.S. Nuclear Regulatory Commission
JERRY PERRIZO, Air Force Office of Scientific Research
ROBERT WESSON, U.S. Geological Survey
JAMES WHITCOMB, National Science Foundation
GEORGE ZANDT, Lawrence Livermore National Laboratory
ARTHUR J. ZEIZEL, Federal Emergency Management Agency

iii

Preface

The Panel on Regional Networks was formed to evaluate the health and status of regional seismograph networks and provide annual reports to the Committee on Seismology that include recommendations for improvements in funding and for operations and research, instrumentation, and networking. Such reports are issued as published documents when appropriate.

Regional networks are supported by federal and state agencies and private enterprise. About 50 regional networks operate autonomously within the United States. The instrumentation for the networks is not standardized and is antiquated, in many cases. Some significant portion of the funding for regional network operation and research is becoming increasingly vulnerable. New thrusts at the federal level for electrical energy production and earthquake monitoring, in general, are resulting in the need to evaluate (1) regional network goals and planned lifetimes of networks; (2) stability of funding for both network operations and research; (3) operational problems, including lack of coordination between regional networks, obsolescence of equipment, rising telemetry costs, poor data quality, and data base management problems; and (4) coordination with national and global networks.

The basic purposes of this report are (1) to make a convincing case for the intrinsic value of regional seismic networks, (2) to describe the seriousness of persistent problems in the current configuration and operation of these networks, and (3) to outline recommendations for their modernization and future evolution, in particular, their short-term integration and long-term affiliation with the U.S. National Seismic Network.

Important supplementary information is included in two appendixes.

Appendix A summarizes results from a survey of regional networks; it provides a snapshot, circa 1989, of the nationwide regional seismic network resources. Appendix C reproduces a valuable, incisive document independently developed by D.W. Simpson, of Lamont-Doherty Geological Observatory, that addresses many of the same issues faced by this panel and offers specific recommendations for the modernization and improved operation of regional seismic networks.

The problems examined by the panel are not new, but some of the options are. Problems associated with regional seismic networks were identified in a prior report of the National Research Council (Committee on Seismology, 1983). Basically, these related to functional definition (the need for a clear statement of network goals and planned lifetimes), funding difficulties, and operational problems (obsolete equipment and the need for standardization and coordination).

By the mid-1980s, future funding for regional networks was critically low, and in October 1985 a symposium and workshop on regional seismic networks were convened in Knoxville, Tennessee, under the auspices of the Committee on Seismology (Simpson and Ellsworth, 1985). More than 100 seismologists attended that meeting, representing the vast majority of the more than 50 regional seismic networks in the United States, and all expressed concern for their future support. A dramatic result of the Knoxville meeting was a ground swell of consensus, enthusiasm, and commitment for addressing in a coordinated way the multifold problems faced by regional networks. The participants unanimously agreed that out-of-date instrumentation was the greatest source of scientific handicap and frustration to network seismologists: handicap, because the type and quality of seismographic data from many regional networks are inadequate for application to current seismological research; frustration, because the technology is readily available to eliminate the handicaps.

Since the Knoxville meeting, a clear consensus has continued to emerge among the seismological community about the urgent need for change—changes in field instrumentation, modes of data transmission, network recording systems, and methods of data analysis and data management. Importantly, the functional objectives of seismic networks have been scrutinized and placed on a firm scientific footing. An initial attempt to do this was made by the Ad Hoc Committee on Regional Networks (ACORN, 1986), appointed at the Knoxville meeting. It has now been done more elaborately in the writing of the "National Seismic System Science Plan" (Heaton et al., 1989) following a July 1987 meeting of federal government and university seismologists in Alta, Utah.

Finally, mindful of how wisdom tends to be "rediscovered," the panel is pleased to point the reader back to the report of the Panel on National, Regional, and Local Seismograph Networks, an earlier Committee on Seismology panel chaired by B.A. Bolt (Committee on Seismology, 1980). In slightly modified form, many of the thoughtful recommendations made by that panel are still relevant to issues affecting regional networks today. The legacy of that panel is captured in the following statement: ". . . the central recommendation of the Panel is that the guiding concept be established of a rationalized and integrated seismic system consisting of regional seismic networks run for crucial regional research and monitoring purposes in tandem with a carefully designed, but sparser, nationwide network of technologically advanced observatories" (Committee on Seismology, 1980, p. 2). Now, 10 years later, the plans, infrastructure, and partial funding for a skeleton national network have been secured, but the precarious status of the nation's regional networks jeopardizes full realization of the powerful tandem system envisioned by the Bolt panel. Since the completion of this study, the Loma Prieta earthquake of October 17, 1989, (with a magnitude of about 7.1) caused damage both in the epicentral region and in vulnerable areas of San Francisco, some 60 miles away. The U.S. Geological Survey operates a regional network in the San Francisco Bay area, and, although the local spacing of sensors was sparse in the epicentral region, this local network provided valuable data on foreshocks, the main shock, and early aftershocks. Combined with the results of portable instruments that were deployed after this quake, these data will help to make Loma Prieta one of the best-analyzed seismic events of the century.

Contents

Executive Summary

Regional seismic networks with centralized recording began in the late 1960s. Without an infusion of new instrumentation, commitment, and funding, most will either cease to exist or be technologically obsolete by the early 1990s, a brief lifetime indeed for such a major observational resource of the geosciences. Are regional seismic networks still necessary? If not, they should be phased out. But if they are, the means must be found not only to continue support for their operation, but also to modernize them so that their important future contributions to basic science and to seismic hazard mitigation can be fully realized. These issues and the various options for addressing them make up the substance of this report.

The threat posed by earthquakes in the United States is actually a mosaic of different problems requiring different approaches to assessment and mitigation. Ours is the only country in the world that must deal with the diverse seismic hazards arising from the full range of earthquake environments, i.e., plate subduction zones (in the Aleutians and the Pacific Northwest), a transform plate boundary (the San Andreas fault in California), hot spots (beneath Hawaii and Yellowstone), distributed plate boundaries (along the Intermontane belt and the Basin and Range province), and major earthquakes of the stable continental interior (New Madrid, Missouri; Charleston, South Carolina). Such diversity presents both major problems in the context of earthquake hazards and major opportunities in terms of understanding the dynamics of the planet.

A concerted national effort to systematically monitor the nation's earthquakes and to gain sufficient understanding to reduce their impact can be achieved.

A principal vehicle for reaching these goals would be a partnership between the U.S. National Seismic Network (USNSN)—planned by the U.S. Geological Survey for implementation in the early 1990s—and a group of streamlined and modernized independently operated regional seismic networks, sited in the important seismic zones of the nation. The combined facilities of the national and regional networks, as proposed in this report, would constitute a National Seismic System, a satellite-based network capable of systematically monitoring and analyzing earthquakes throughout the nation within minutes of their occurrence. Such a system would maintain the vital regional research and response flexibility required by our nation's diverse seismic zones, and its dual components each would have significantly increased capabilities beyond those possible in isolation. Clearly, a National Seismic System can be a whole greater than the sum of its parts.

The USNSN is designed and intended to detect and report on only those U.S. earthquakes above magnitude 2.5-3.0; research considerations are secondary to this mainly operational intent. The addition of the regional network component to form a National Seismic System would expand USNSN's capability to be a national research facility of unprecedented effectiveness. Also, it is important to note that regional networks supply a continuity of seismicity data essential for seismic hazard evaluation, short-term earthquake forecasting, or even longer-range predicting. These data must be in place when the need arises—they cannot be gathered after the fact.

Two examples illustrate how the regional networks will augment the USNSN. First, the wide-aperture USNSN can provide three-dimensional locations of earthquake foci to within about ±5-10 km; dense regional networks can improve this to ±1-3 km for earthquakes in their area. Over much of the United States, crustal faults capable of producing damaging earthquakes have minimum dimensions of less than 10 km. Only regional networks with closely spaced stations and microearthquake detection capability have the resolving power necessary to delineate such features.

Second, the powerful technique of seismic tomography developed during the 1980s is dependent on dense sampling of the earth's crust by seismic rays. Just as medical CAT-scans provide the surgeon with three-dimensional images of the interior of the human body, so also does seismic tomography provide the seismologist with three-dimensional images of the geologic structure of the earth's interior. Such high-resolution images are fundamental to achieving advances in understanding and dealing with all earth processes, including earthquakes.

The USNSN, with an average station spacing of about 370 km, cannot adequately resolve the details of shallow earthquakes within the continental crust. Crustal tomography will require the operation of special arrays or the continued operation of regional seismic networks, which have the advantage of providing long-term recording. The panel considers that seismic tomography is a technique of such great promise that enabling its use alone justifies the operation and upgrading of regional seismic networks. Indeed, it is mainly because of the potential scientific gains afforded by seismic tomographic investigations that the panel foresees the need for an increase in the number of regional seismic network stations rather than the pending decimation that will result from withdrawal of seismic network support by federal agencies, principally the U.S. Nuclear Regulatory Commission.

The current deployment of regional seismic networks in the United States is outlined in Appendix A. Nearly 50 organizations operate about 1,500 seismograph stations (roughly 40% of which are sited in California). Because of inadequate finances, fewer than 10% of the 1,500 stations record complete seismic waveforms, and fewer than 3% incorporate state-of-the-art design in their electronics. The panel recommends that a concerted program of regional network modernization be a high-priority objective of the proposed National Seismic System.

A National Seismic System with the USNSN forming the backbone or framework would have operational advantages. The data communications, data management, and data distribution systems of the national network could be used by the regional networks. The regional networks in turn could provide local support for national stations within each region. The result would be greater efficiency in operations on both sides and more standardization in data collection, production of routine data-based products, and generation of software, thus making data exchange between networks easier. In addition to these tangible benefits, a National Seismic System would allow seismologists from both the regional or the national perspectives to speak and act from a stronger, more unified position.

The quality and scope of both the national and the regional components of a National Seismic System will be controlled by financial considerations. For this reason the panel, in its "Findings and Recommendations" (Chapter 7), recommends a modest increase in the projected funding for the proposed National Seismic System and an increase in support for network operations from the current level of approximately $10 million per year from diverse sources to $12 million per year. Additionally, the panel recommends a one-time capital investment of $15 million spread over a five-year period. The $12 million incorporates the funds necessary to operate a complete National Seismic System and to continue the operation of regional networks in the principal seismic zones of the nation. The $15 million represents the funds necessary to (1) expand the USNSN from only the eastern United States to the entire nation, (2) provide satellite data links between the national center and regional network operation centers, and (3) provide for the needed gradual upgrading of regional network instrumentation and recording facilities.

The recommended increase from $10 million to $12 million per year for operating a National Seismic System and the $15 million for capitalization and modernization constitute funding that is modest considering the cost-benefit ratio. The proposed National Seismic System is an idea whose time has come. It should be fully implemented without delay. With it seismology can take a major step forward in the fundamental study of planet earth and in the determination of the earthquake hazard to which Americans are subject.

1

Introduction and Background

Regional seismic networks are discrete arrays of tens to hundreds of seismic stations targeted chiefly on seismically active regions. They are a fundamental, multipurpose tool of observational seismology, providing a broad range of data and information. Data acquired by these networks have a host of applications, including but not restricted to public safety and emergency management; quantification of hazards and risk associated with both natural and human-induced earthquakes; surveillance of underground nuclear explosions; and wide-ranging basic research encompassing earthquake mechanics and dynamics, seismic wave propagation, seismotectonic processes, earthquake forecasting and prediction, and properties and composition of the crust and of the deeper internal structure of the earth (for a comprehensive overview, see Heaton et al., 1989). Importantly, regional seismic network facilities are also essential for the graduate education and training of this country's professional seismologists, and they provide the most readily available sources for public information and for expert assistance to public policymakers, planners, designers, engineers, and safety officials on the local and regional level.

Previous National Research Council reports (Committee on Seismology, 1980, 1983) have distinguished regional from local seismic networks on the basis of scale, lifetime, and mission. In these reports, as in this one, "network" means "a collection of seismic stations operated coherently, normally by one organization, with a common basis for data collection and analysis" (and typically with telemetry to a central recording and analysis facility). Local networks are characterized by dimensions smaller than several tens of kilometers, an operating lifetime of less than several years, and a specialized research and monitoring mission focused, for example, on a critical facility (such as a dam or nuclear power plant) or a localized seismic source zone (such as a volcano or geothermal area). Local networks are often operated by private companies.

Regional networks operate on a scale ranging from hundreds of kilometers to 1,000 km. They have an unspecified lifetime, but are commonly assumed to be permanent facilities, and they are generally operated by government agencies or universities. Figure 1 gives an overview of three fundamental aspects of the role or mission of a regional seismic network (note that the three functions are not mutually exclusive or in order of priority): earthquake monitoring and rapid emergency response; scientific research; and the acquisition of information required for earthquake hazard and risk analyses as well as for earthquake engineering. Efforts aimed at earthquake forecasting and prediction apply to all three functions. Thus regional networks play an essential, if unrecognized, role far beyond that of simply monitoring earthquake activity.

Currently, there are about 1,500 seismic stations operating in the United States, forming parts of about 50 regional seismic networks (Appendix A). Figure 2 shows the distribution of these stations, some of which may be construed to be part of local networks. Because the panel supports the goal of improving network seismology in the United States, it has not arbitrarily excluded all consideration of local networks. Nevertheless, the panel's recommendations chiefly address regional seismic networks as defined above.

The vast majority of current regional seismic network instruments are substandard when compared with the needs of modern seismological practice (see Appendix A). Specifically, they consist largely of vertical-motion-only sensors, recorded over a narrow frequency band (~1-20 Hz) with limited dynamic range (~40-60 dB). The desired operational characteristics of a modern network would include full three-component recording with a much higher dynamic range (>100 dB) and with at least a subset of broadband stations. Not only is there no plan to modernize these networks to achieve their full potential, but instead decreasing federal operating support is eroding their capabilities.

The panel has found a crisis atmosphere affecting regional networks nationwide. The decision of the U.S. Nuclear Regulatory Commission to phase out its support of regional networks in the eastern United States and to support instead the development of a U.S. National Seismic Network (USNSN) by the U.S. Geological Survey (USGS) has already begun to curtail network operation and student involvement (see Appendix B). In the western United States, both federal and federally supported networks are suffering because of inflation-eroded, no-growth funding of the USGS budget for the last six years.

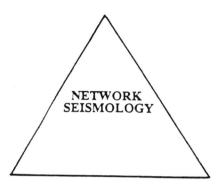

Function: Scientific Research
Users: (Scientists and Engineers)

NETWORK
SEISMOLOGY

Function: Earthquake Monitoring & Rapid Emergency Response
Users: (Public Safety Officials, News Media & General Public)

Function: Input to Earthquake Hazard & Risk Analyses, Earthquake Engineering
Users: (Engineers, Public Officials & Other Decision Makers)

- Earthquake data base
- Seismotectonic framework
- Earthquake source identification
- Seismicity parameters & earthquake occurrence modeling
- Information for predicting strong ground motion (source mechanics, attenuation)

Figure 1. The multifold practical functions of regional seismic networks.

Because large damaging earthquakes in the United States are episodic, public attention and concern wax and wane, and the potential of earthquakes to cause great sudden disasters is often ignored. As a result, earthquake seismologists have been unable to gain adequate sustained support from representatives and officials charged with taking a long-term view on society's behalf. All of the major infusions of funds that have enabled seismology, including regional networks, to grow have been the result of specific missions, rather than a fundamental national commitment to the science. The major missions—e.g., nuclear test monitoring and the assessment of earthquake

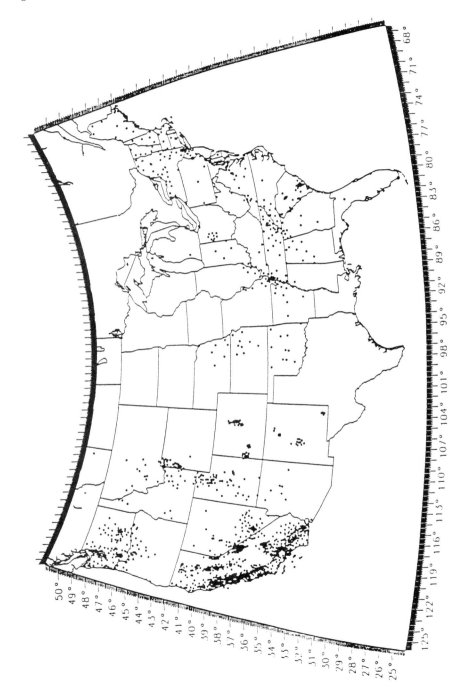

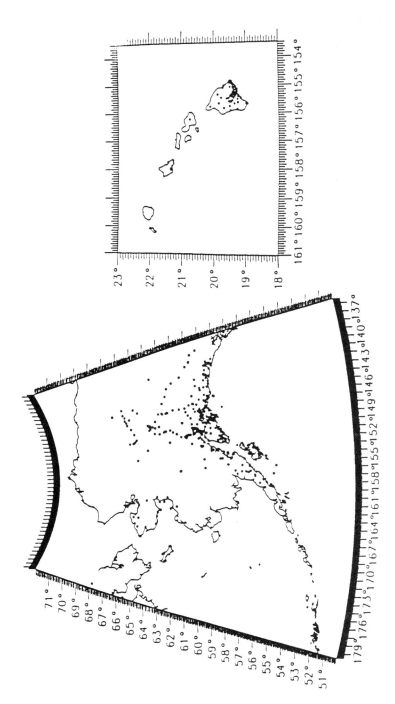

9

Figure 2. Regional seismic network station distribution in the United States, 1989 (modified from Heaton et al., 1989).

hazards for siting of critical facilities, earthquake prediction, and hazardous waste disposal—have come at irregular intervals. The result has been the lack of stability in support for regional networks, to a degree unique to the United States among technologically advanced nations.

The U.S. seismological community is making coordinated efforts to modernize and streamline the capabilities and effectiveness of regional seismic networks. These efforts presuppose that stable, long-term funding can be secured from funding agencies, when policymakers are convinced of the importance and value of such investment. Only a modest amount of sustained support is required (see Chapter 7).

The development of the USNSN (see Chapter 5) has contributed to the regional network crisis, as noted above. Funding for the limited deployment of the USNSN has undercut and will soon eliminate the support currently received by most of the central and eastern U.S. regional networks. The sparse station spacing of the USNSN, however, means that many fewer earthquakes will be recorded and that for those that are, the locations will be determined with less accuracy than is possible when using regional network data. Is such detailed information still needed, or are regional seismic networks obsolete? This report is intended to answer that question. The panel finds that the USNSN is essential to the nation's need for information about earthquakes but that it is, by itself, insufficient to provide all of the needed information. The panel also finds that the regional seismic networks have been an inefficient means of producing the needed information because they are regional and isolated and operate without adequate facilities and support staff and without a unifying national support system.

The following chapters of this report present the basis for these findings. They deal with the contributions to date of regional seismic networks (Chapter 2), problems and limitations of the networks in their present form (Chapter 3), the case for a continuance of regional network operation (Chapter 4), and a specific initiative that holds great promise for revitalizing regional networks (Chapter 5). The panel envisions an essential and productive future for regional networks as an integral partner with the developing U.S. National Seismic Network (Chapter 6). After a brief look at this future, the report ends with a set of specific recommendations (Chapter 7), which if followed, could make this partnership—a National Seismic System—a reality.

2

Contributions to Date of
Regional Seismic Networks

Most regional seismic networks currently in operation in this country have been sited to monitor active seismic zones. Because they consist of multiple sensors distributed over relatively small areas, they are essentially telescopes focused downward into the earth to "see" the seismic source. Such networks have been in operation for only about two decades but have made extensive contributions to our knowledge of the spatial, temporal, and physical characteristics of earthquake occurrences. Heaton et al. (1989) recently reviewed these contributions and discussed the future of networks in the context of a science plan for a National Seismic System. Briefly, the contributions include the improved detection and more accurate location of earthquakes, especially those of lower energy levels; greater precision in focal depth determinations; enhanced monitoring of seismic energy release as a function of space and time; refined determinations of the attenuation characteristics of seismic waves; three-dimensional descriptions of the seismic velocity structure of the interior of the earth; and more reliable specification of the earthquake faulting process. Thus the fundamental contributions from seismic networks are intrinsically observational, and these observational data make possible a wide range of derived contributions that are of direct benefit to both science and society.

Recent examples of such contributions with direct societal benefits are described in two earthquake case studies in Chapter 4. Other examples include contributions from networks associated with active volcanoes such as Mount St. Helens in Washington and Kilauea in Hawaii. The networks there track the subsurface motions of magma bodies and their associated mechanical deformation and thereby provide invaluable data to emergency preparedness agencies. Similar contributions are made when damaging earthquakes occur in populated areas, such as happened in one of the case study events, the Whittier Narrows shock in the Los Angeles basin. In addition, data from regional seismic networks are essential to the safe siting of nuclear and other hazardous waste repositories as well as large, unique engineering structures such as the proposed Superconducting Super Collider. Siting such structures safely requires an already-developed adequate seismicity data base; once a site has been proposed, it is not possible to wait for data to be gathered.

Seismic networks provide a major contribution to the estimation of U.S. seismic hazards, which vary greatly across the nation: seismicity is highest on the West Coast, but 37 states are in the two highest (out of four) risk zones. The current federal National Earthquake Hazard Reduction Program (NEHRP) recognizes this pervasive threat and seeks to mitigate it. The program was created by the Earthquake Hazards Reduction Act of 1977; its principal agencies are the Federal Emergency Management Agency, the U.S. Geological Survey, the National Science Foundation, and the National Institute of Standards and Technology. One of the major NEHRP elements is "hazards delineation and assessment" (FEMA, 1988). In particular, the estimation of seismic hazard requires as input (1) spatial definitions of the seismic source zones (especially important is the accurate definition of currently active geologic structures as well as their seismotectonic host environment, e.g., the thickness of the crustal seismogenic zone); (2) rates of occurrence of earthquakes of various magnitudes for each zone; and (3) ground motion estimation for seismic vibrations from earthquakes of varying magnitudes and at varying distances. Clearly, only the highly accurate data from dense regional seismic networks that are dedicated to the investigation of specific seismic zones or regions can provide adequately for such specific requirements. This is especially true for the eastern United States, where the seismic station density before networks were established in the 1960s and 1970s was lower than one per state. It is important to reemphasize that the required input data from the regional networks cannot be obtained as the need arises for their use; rather they must be obtained before such needs arise. It is also important to note that the determination of seismic risk—i.e., the estimation of probable consequences of earthquakes for life and property—depends directly on the availability of accurate seismic hazard estimations, which in turn are based largely on data from regional networks.

The technological growth of industry in this country, in concert with increased land use during the past several decades, has resulted in a dramatic increase in the elements of society at risk from earthquakes. Engineers have constructed larger and more complex facilities, such as long bridges, high dams, high-rise buildings, nuclear reactors, large computer centers, offshore drilling platforms, and concentrations of high-technology industry. These and other critical facilities are often sited in areas of high population density that are also earthquake-prone, e.g., the computer chip industry in California. In addition to estimating the seismic hazard for such facilities, it is also necessary to thoroughly evaluate the probable responses of the structures themselves to seismic disturbances. Such studies are based directly on the best possible estimations of the amplitudes and frequencies of ground motions from both moderate and large earthquakes at distances ranging from nearby to regional. Seismic networks, especially those that include strong-motion seismographs sited in the structures themselves as well as in the free field, are the only source of the input data required for the necessary estimations (Committee on Seismology, 1980). Clearly, a lack of such monitoring efforts exposes our society to increasingly unacceptable and unspecified risks from future earthquakes.

One of the current frontiers of research in seismology involves the prediction of earthquakes. For example, the U.S. Geological Survey has predicted that a magnitude* 6 earthquake will occur on the Parkfield section of the San Andreas fault in 1988 ±5 years. In general, however, the present stage of development of this research field is such that estimations of future earthquake occurrences are generally derived in more probabilistic terms and are based on detailed analyses of the spatial and temporal patterns of earthquake activity in the forecast area. Both probabilistic and deterministic analyses include the recognition of (1) "seismic gaps," i.e., locales that are known from prior activity to be earthquake-prone but currently are seismically quiescent, (2) repetitive "characteristic" earthquakes from a given fault segment, and (3) "slip-deficient" fault segments. Only the resolving power of the inward-looking regional seismic network "telescope" can provide data of adequate precision, detail, and completeness to satisfy the requirements of this most difficult and demanding seismological task—that of predicting earthquakes in a quantitative manner. However, the benefits to society that would result from this ability are so enormous that we must continue these efforts.

The dense spatial coverage provided by regional seismic networks has been directly exploited in recent studies of crustal velocity structure. Some of these studies are similar in concept to computer-assisted X-ray tomography, the CAT-scan in medical technology, which yields three-dimensional, computer-generated "images" of the interior of a body without directly accessing the volume being investigated. For example, Hearn and Clayton (1986) have presented detailed images of lateral variations in the shallow crustal

---

*"Magnitude" as used throughout this report is a generic term for the relative size of the earthquakes discussed. The term may refer variously to a body wave, surface wave, moment, 1-g, local, or Richter scale magnitude.

velocity structure in southern California that they obtained from data derived from the seismic network stations located there. These velocity variations are associated with surface tectonic features such as the San Andreas fault. Also in southern California, Humphreys et al. (1984) studied the deeper mantle structure beneath the Transverse Ranges to image a spectacular, high-velocity tabular root extending several hundred kilometers into the mantle (see also Heaton et al., 1989). In the Midwest, Al-Shukri and Mitchell (1988) mapped a three-dimensional pattern of low velocities in the crustal rocks of the active portions of the New Madrid fault system in southeastern Missouri. The seismic velocities there are lowest in those subsurface volumes exhibiting the greatest concentration of earthquake activity. The observed several percent decrease in compressional wave velocity is consistent with a source zone containing fluid-filled cracks. Studies such as the three mentioned here have led to a markedly improved understanding of the physics and architecture of the earth's crust. Again, the many stations of the regional seismic networks are required to achieve the detail and resolution necessary to accomplish such CAT-scans of the earth.

When large fault motions occur on the floors of oceans, they produce not only earthquake vibrations but also energetic water waves, called tsunamis, that travel across the oceans and run up on distant coastlines. Between 500,000 and 1 million residents along the coastlines of Hawaii, California, Oregon, Washington, Alaska, and the U.S. Pacific Territories are at risk from these rare but devastating waves. For example, the 1964 Alaskan earthquake (magnitude 9.2) generated a tsunami that caused 122 fatalities in Alaska, California, and Oregon and $100 million in damage in Alaska, Hawaii, California, Oregon, and Canada. Tsunamis are predominantly, but not exclusively, a Pacific hazard: in the Atlantic Ocean in 1929, the Grand Banks earthquake off the coast of Newfoundland (magnitude 7.2) also resulted in damage and fatalities (Committee on Seismology, 1980; Lander and Lockridge, 1989). Additionally, submarine facilities, such as communications cables, are at risk from these earthquakes as well as from submarine landslides triggered by earthquakes. The Pacific Tsunami Warning Center at Honolulu, Hawaii, is an international cooperative effort to provide tsunami watches and warnings to the Pacific region. Onshore regional seismic networks contribute to the detection and location of submarine earthquakes that are potentially tsunamigenic. Needed, but not currently in place, are networks of ocean-bottom seismographs on U.S. continental shelves to improve detection and location capabilities in those nearshore areas. The combined onshore and ocean-bottom seismic networks would allow for a more rapid determination of focal mechanism and thus a more reliable assessment of the tsunami-generating potential of shallow offshore events.

Earthquakes are common in volcanic areas, and seismic networks are the fundamental tool for their study. Data from networks have shown that "volcanic earthquakes," those that result from the thermal and mechanical forces of volcanic processes (volcanic a-type, high-frequency earthquakes), are indistinguishable from tectonic earthquakes, which result from the mechanical fracturing of rock due to tectonic forces. Other volcanic earthquakes (volcanic b-type, low-frequency events) and harmonic tremor (vibrations due to the shallow movements of magma) have distinctive properties. For example, studies at Mount St. Helens indicate that harmonic tremor there consists of a persistent sequence of b-type earthquakes. Studies in Hawaii and Alaska have resulted in the development of new models of the sources for the volcanic shocks that include reverberations within the magma body triggered by brittle failure of the adjacent rock as well as a point-force reaction to an explosive eruption. The swarm-like series of magnitude 5.5-6.0 earthquakes that occurred in 1978 near the Long Valley caldera in eastern California raises the possibility of yet another type of volcanic earthquake, one due either to the abrupt injection of magma into a dike or to a complex shear failure on fault planes of differing orientations (Hill, 1987). Clearly, much work remains to be done to understand what the various types of volcanic earthquakes imply about the volcanic processes that affect the westernmost states.

The core of the earth has long held a particular fascination and position of importance because of its inaccessibility and because it is the source of the earth's magnetic field. Regional networks, when integrated within a continent-wide National Seismic System, can contribute to its study. Recent studies of the structure of the core and of its boundary with the mantle using compressional waves that penetrate through the deep interior of the earth suggest considerable complexity that could have important geodynamical and geochemical consequences. It appears that topography of ±8 km or so may be present on the core-mantle boundary. Establishing whether that boundary is thermal or chemical in nature is important for thermal modeling of the earth's interior. Also, although the velocity gradients in the outermost core appear not to be anomalous (as was once thought), and although the inner core-outer core boundary may indeed be a simple discontinuity, the first-generation three-dimensional core models indicate greater, not less, complexity for core structure (Lay, 1987). The rapid progress made in imaging these most inaccessible regions testifies to the benefits that can be reaped from the high-quality data derivable from the larger regional and global networks.

Finally, the importance of seismological facilities for education deserves emphasis. This includes not only the training of the nation's seismologists but also the general education of a broad student population. Terminating funding for some seismic networks will cause a certain number of researchers to seek new avenues of funding in more adequately supported areas of research. Once these scientists are lost to other research fields, they cannot easily be reclaimed for seismic network studies even if funding priorities change. Thus, given the small number of network seismologists to begin with, a short-term reduction of support will have long-lasting consequences.

Students at universities that operate regional seismic networks unquestionably have an enhanced educational experience. The incoming digital data stream from multiple sensors provides hands-on opportunities to apply and develop the seismological theories developed in the lecture hall and the laboratory. Not only can near-real-time analyses be performed, but the presence of a continually expanding archival digital data base also permits a full range of thesis and dissertation investigations. The day-by-day, real-time acquisition of seismic data provides an earth surveillance setting and format that are particularly dynamic and impart to students an excitement about earth processes that often lasts a lifetime.

In summary, regional seismic networks have made fundamental contributions to the estimation of national seismic hazards and strong earthquake ground motions, the prediction and forecasting of earthquakes, the specification of the three-dimensional internal structure of the earth, the surveillance for tsunamis, the study of volcanic earthquakes, and the training of students. Such worthwhile efforts should be continued and enhanced.

3

Problems and Constraints

Regional seismic networks have been faced with numerous funding and operational challenges virtually since their inception. Many of these were documented in the report of a 1982 National Research Council workshop (Committee on Seismology, 1983). The problems have become more acute as the network instruments age and as funding drops. The principal problems arise from three causes: (1) obsolete, aged, and narrow-focus instrumentation; (2) difficulties in handling large volumes of network data; and (3) labor-intensive operations with significant capitalization requirements.

OBSOLETE INSTRUMENTATION

Obsolete instrumentation is a major problem facing the regional seismic networks. Many powerful new analytical techniques, developed over the last 10-20 years, require higher-quality data than current regional networks can supply. An overwhelming majority of the more than 1,500 stations in existing regional seismic networks consist of short-period, vertical seismometers that were developed and installed one to two decades ago. The FM radio telemetry system used to transmit nearly all the data was developed over 25 years ago; the resulting signals have a narrow frequency band (~1-20 Hz) and low dynamic range (often only 40 dB). This type of system produces seismic signals with clear P-wave arrivals, well suited to the task of locating and determining first-motion focal mechanisms of local earthquakes in a relatively effective manner. (Focal mechanism determination is a technique by which the orientation (strike and dip) of a fault and the direction of slip on that fault are determined from the radiation pattern of seismic waves generated by the earthquake and recorded at numerous seismic stations.) The response and sensitivity typical of a regional network seismograph, relative to representative levels of earth noise and earthquake ground accelerations, are illustrated in Figure 3. The typical station—producing one-component, narrow-band, low-dynamic-range data—is inadequate for studies of moderate to strong ground motion, teleseismic earthquakes, or rigorous waveform analysis; and most importantly, the signals are usually "clipped" (i.e., the recording system is overdriven) for local earthquakes of magnitude 4, regional events of magnitude 5, and teleseisms of magnitude 6-7.

Even for the purpose of earthquake monitoring, for which many of the regional seismic networks were installed, existing instrumentation is inadequate for some routine earthquake cataloging tasks. Determination of magnitudes over the normally recorded range $(1.0 < M < 6.5)$ is often impossible because of the limited dynamic range of the sensing-recording system. The accuracy of depth determinations of earthquakes is greatly improved if S-wave arrivals are included, but they are poorly recorded by vertical seismometers, which are the only sensor component deployed at the great majority of stations. Thus, the instrumentation of regional seismic networks, while relatively inexpensive in initial cost per station, ultimately has penalized regional networks in terms of missed research opportunities. Unfortunately, the relatively unsophisticated instrumentation has served to isolate regional network operations from the forefront of the seismological community, which relies on the advanced technology of relatively few stations for state-of-the-art analyses.

DATA-HANDLING DIFFICULTIES

Handling large volumes of network data poses additional problems. The large number of stations and low-magnitude threshold of regional networks lead to such a large quantity of seismic data that only computer-based storage and manipulation of the data are feasible. Although the use of computers for the acquisition, processing, and storage of data has become standard for regional networks, the computer systems and software used have not. Different individually developed and generally undocumented systems are in use at different networks, which makes internetwork data exchange difficult. It is obviously inefficient for each individual network operation to develop its own software for data analysis. Because written documentation for the systems is commonly lacking, use of the data by visiting scientists often demands significant time from the network operator or data analyst to explain the local system. This difficulty in accessing data has restricted the usefulness of regional networks. In most cases, the fundamental raison

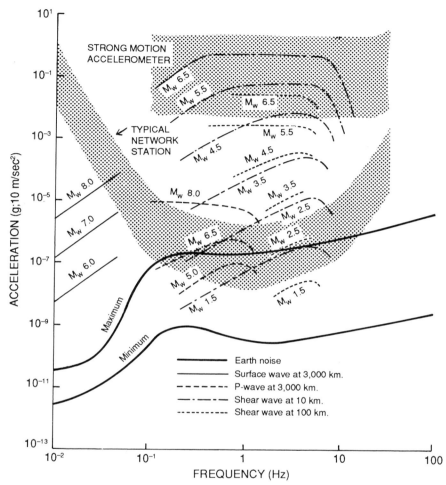

Figure 3. Dynamic range versus frequency for a typical station of a regional seismic network and a typical strong motion compared with the expected levels of ground motion (acceleration) for seismic waves from earthquakes of differing size and distance (from Heaton et al., 1989). For the ground acceleration, the units have been approximated to 1 g = m/s². M_w is seismic moment magnitude and may be taken as equivalent to the generic magnitudes used in the text. (From Heaton et al., 1989.)

d'être of the regional network is the generation of a local earthquake catalog, and this objective is almost always fulfilled. However, the scientific objective of furthering the understanding of local seismotectonic structure and earthquake hazards requires an in-depth analysis by scientists with a variety of backgrounds. In the worst case, data inaccessibility means that such worthwhile studies are never undertaken.

OPERATIONAL DIFFICULTIES

The operation of regional networks is especially time-consuming both for scientists managing the operation of the network and for analysts processing the data. The close involvement of research seismologists with the operation of a regional network is essential to maintain the integrity and usefulness of the network as a scientific tool. For the seismologists, however, this represents a drain on time that usually detracts from research time. Although the situation has improved with the use of computers, the processing of earthquake data requires much meticulous, albeit repetitive, work by data analysts. During times of budget constraint, these manpower requirements are sometimes not met, forcing strict economy measures, which in the worst case can lead to unprocessed, or even lost, data.

ARE REGIONAL NETWORKS COST-EFFECTIVE?

Another problem faced by operators of regional seismic networks is the perception that the costs of their network operations are high—at least compared to budgets typically prepared by academic seismologists for competitive research funding. What are these costs? Can they be reduced significantly? Are large networks more economical than small ones, and are university-operated networks more costly than federally operated ones? Care has to be taken in addressing these questions because there are evident pitfalls, especially in comparing costs reported for individual networks. Available surveys of cost information—including the one conducted for this report (see Appendix A)—do not contain uniform or complete information. Seismologists are not experienced accountants, and hidden expenses such as the cost of facilities, complete personnel costs, benefits, and separately paid telemetry or computer maintenance charges may be unintentionally neglected.

Despite their recognized shortcomings, surveys of network operators remain the best source of information about true costs at the individual network level. Figure 4 shows the annual operating costs reported to the panel by the operators of 43 regional (and local) seismic networks in the United States versus the number of stations in the network. Operators were asked to report "total number of stations" and "total current annual funding for network operations alone, exclusive of research." There may be a differ-

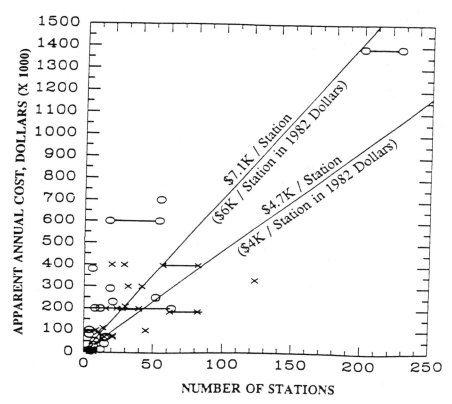

Figure 4. Apparent annual cost versus number of stations for 43 regional (and local) seismic networks in the United States. X's indicate university-operated networks; O's, nonuniversity networks. Tie lines give the range where the number of stations operated by a network (the smaller number) differs from the number of stations recorded. The information comes from a survey conducted for this report in late 1988 by the panel. Trend lines for cost per station relate to a report on seismic networks by the Committee on Seismology (1983; see text). The values of $7.1K per station and $4.7K per station are adjustments to 1988 dollars from originally reported values of $6K per station and $4K per station, respectively.

ence between the number of stations recorded (in all cases less than or equal to the total number of channels recorded) and the number of stations actually maintained and operated by a unit. Where a network operator has distinguished between the numbers of stations operated and recorded, Figure 4 shows both data points connected with a tie line. The two largest regional seismic networks in the United States are CALNET, a network of 327 stations operated by the USGS in central California (operational costs for CALNET were not provided to the panel), and the Southern California Seismic Network of more than 200 stations, jointly operated by the USGS and the California Institute of Technology.

In a 1982 survey made for the Committee on Seismology (Committee on Seismology, 1983), trend lines for cost per station of $4,000 and $6,000 enveloped average survey results. These trend lines, adjusted to 1988 dollars, are superimposed on Figure 4 because the dollar estimates still tend to be cited in debates about network costs, although as is apparent in Figure 4, the relation of operational costs to number of network stations is not simply linear.

The data summarized in Figure 4 show that the annual operational cost for a seismic network that is truly regional in character is on the order of hundreds of thousands of dollars and exceeds a million dollars for the largest networks. The scatter evident for annual operational costs as a function of network size is remarkable and is partly explainable by factors already described. Despite the scatter, the information provides some useful insights. First, the large California networks that have more than 200 stations should probably be considered in a category by themselves, with distinctive costs and economies of scale. At the other end of the scale, Figure 4 shows a distinctive grouping of 22 networks (close to half the survey sample) having 21 or fewer stations and an annual reported cost of $110,000 or less. This group appears chiefly to include small networks surviving on minimal funding and solidly established networks whose true total costs, arguably, may not have been completely accounted for. Except for two networks whose apparent annual cost is $600,000 or more, the remaining 18 networks characterized in Figure 4 seem to define a pattern marked by annual operational costs of approximately $200,000 to $400,000—despite having numbers of stations ranging from less than 20 to 123. Because capital costs for station hardware are not included in this analysis, the $200,000 to $400,000 cost range appears to reflect a fundamental threshold of operational costs for regional networks of moderate size.

Are these operational costs (exclusive of research) excessive? A simple analysis, using for convenience the federal pay scale for FY 1989 as a reasonable index of salaries, may suggest an answer.

A generic regional network of, say, 50 stations might conservatively require the following: one quarter-time equivalent for management by a Ph.D.-level seismologist with three years' experience (GS-13 or equivalent); one full-time, M.S.-level seismologist (GS-10); one full-time field technician (GS-8); one full-time seismic analyst (GS-5); and one full-time secretary (GS-5). The resulting total for annual salaries would be $117,000. (The GS-ratings used for this analysis were intentionally pegged low for the sake of argument. In some networks, such as in the eastern United States, where seismicity is relatively low, student assistants commonly replace the seismic analyst.)

A complete cost profile for a hypothetical, relatively low-cost network might be as follows:

Item	Annual Cost
• Salaries	$117,000
• Employee benefits (at 30%)	35,000
• Telemetry	25,000
• Computer-related costs	20,000
• Supplies	15,000
• Field travel	10,000
• Indirect costs (25%)	55,000
Total	$277,000

The telemetry costs in this example would be for telephone and/or microwave charges. (Actual telemetry costs for some moderate-sized networks in the eastern United States approach $100,000 per year.) Computer-related costs are chiefly for maintenance contracts. Indirect costs arbitrarily have been assigned at the low level of 25% of the total direct costs. (This is roughly half the typical federally approved rate for a university, but the number may be realistic if only a part of the total operational costs comes from federal awards or if some of the costs are paid directly by a federal agency.)

The example above readily shows why annual operational costs for a regional seismic network amount to hundreds of thousands of dollars— whether or not those costs are fully identified. Realistically higher salary levels, greater telemetry costs, greater costs for network maintenance in environments harsher than those in California, and other justifiable factors easily escalate the total costs. Personnel costs for minimal core staffing indicate much about why an average cost per station will be predictably nonlinear for most moderate-sized networks. Finally, Figure 4 shows that university-operated networks do not tend to be more costly than nonuniversity networks.

The panel emphasizes that this example includes no funding for permanent equipment. For most networks, meeting unavoidable operational costs in the face of inflation-eroded, level federal funding has allowed only minimal spending for permanent equipment in the last five years. The modernization of existing seismic networks will involve costs on the order of tens of thousands of dollars per individual seismic station, and a few hundreds of thousands of dollars for the computer-recording-and-analysis laboratories of individual networks. (Survey results summarized in Table A1, Appendix A, highlight the problem of aging computers.) The panel has estimated the cost to modernize the recording centers and a subset of stations of the regional networks and included these estimates as Recommendation 6 in Chapter 7.

In sum, the operation of regional seismic networks involves unavoidable inherent costs that require sustained support. Regional seismic networks are fundamentally wide-area communication networks requiring complex electronics, all-weather remote field installations, telemetry systems for continuous data transmission, elaborate central-recording laboratories with dedicated computers and peripherals for recording and data processing, and well-trained scientists, technicians, and data analysts for efficient and productive operation. The returns for such an investment of manpower and resources have been amply demonstrated in Chapters 1 and 2, and additional benefits to science and society are explored in Chapter 4. The panel strongly believes that when the costs and benefits of regional networks are assessed comprehensively, the latter clearly outweigh the former. The panel thus concludes that regional seismic networks are a cost-effective investment for the nation.

4

The Need for
Regional Seismic Networks

The recognized problems with regional seismic networks examined in Chapter 3 raise the question of whether the new U.S. National Seismic Network (USNSN) would be a superior system for assessing the nation's earthquakes. Both approaches have advantages and disadvantages. The USNSN will open important new avenues for seismological research; by itself, however, it will be insufficient to meet the nation's seismic data needs. The seismic stations of the USNSN are planned to be of very high quality (see Chapter 5), but the number of stations deployed will be small—only 100 to 150 USNSN stations are proposed compared to the more than 1,500 stations now deployed in existing regional networks. High station density is an unavoidable prerequisite to successful analysis in many types of seismological investigations.

One of the important advantages of a dense network of stations is that many more small earthquakes can be detected and located. Because small earthquakes occur much more frequently than large ones and are associated with active tectonic structures, dense networks can define and resolve the dimensions and characteristics of these structures in a relatively short time. For instance, in California, the proposed station spacing of the USNSN will allow determining locations for most earthquakes of magnitude 3.0 or greater, or a few hundred events in an average year. The existing regional networks in California record all earthquakes above magnitude 1.5, or more than 20,000 events every year.

Accurately locating large numbers of earthquakes is important for recognizing and defining active faults and understanding the seismotectonic structure of actively deforming regions (see the Whittier Narrows case study below). Dense regional networks are also necessary for well-constrained determinations of earthquake focal depths and focal mechanisms for events below magnitude 5. These three types of fundamental data—earthquake locations, focal depths, and focal mechanisms—which for many earthquakes are available only from dense regional networks, are essential to many types of seismic studies, including seismic zonation, characterization of source mechanics, and earthquake prediction. Examples of such studies were described in Chapter 2.

With the recent advent of automated, real-time event location and event analysis using computers, regional networks have been able to locate and determine magnitudes for earthquakes within a few minutes of their occurrence. This capability has greatly increased the usefulness of regional networks to emergency management personnel. The possibility that some earthquakes could be foreshocks of larger main shocks has led the U.S. Geological Survey to issue several short-term earthquake warnings, based on data from dense regional networks (e.g., Heaton et al., 1989; Goltz, 1985). Data from dense regional networks have also been crucial to aftershock studies and research aimed at understanding the rupture process of earthquakes.

As was stressed in Chapter 3, a primary limitation of regional networks is the bottleneck created by conventional short-period seismometers linked by analog telemetry, a system that severely restricts seismic wave recording in terms of both frequency bandwidth and dynamic range of amplitudes. Integration with the USNSN would greatly improve the digital telemetry capabilities of regional networks and make it possible at some sites to deploy three-component sensors (two horizontal components in addition to the standard vertical component) with enhanced bandwidth and dynamic range. The special contributions envisioned as coming from such improved regional networks as part of a National Seismic System are described in detail in Heaton et al. (1989).

One program objective that generates considerable interest is short-term warning of imminent ground shaking. In great earthquakes that occur on very long faults, substantial damage is often experienced at large distances from the earthquake's initial rupture point. Because seismic waves travel slowly in relation to electromagnetic waves, it is possible to warn of imminent strong ground shaking from an earthquake that has already started by using electronic messages that can arrive several tens of seconds before the strong shaking. Upon receipt of these messages, local computer systems could trigger automatic safety responses and warnings. Heaton (1985) estimated that such a system could have provided the Los Angeles region more than a minute's warning before the great Fort Tejon earthquake of 1857, an event with a significant probability of recurrence in the next several decades.

Upgraded dense regional networks could play an important role in post-earthquake disaster response and recovery. Rapid estimates of the areas of maximum ground shaking greatly facilitate search and rescue operations immediately following a damaging earthquake. If sensors have sufficient dynamic range to record on-scale both the strong ground shaking during a large earthquake and the weak ground motions during smaller events, data from the much more common small earthquakes can be used to estimate ground response during the large events. This capability, also useful in earthquake engineering and land use planning, currently is severely lacking in the United States.

As discussed elsewhere in this report, the scientific uses of high-quality, dense regional networks are numerous and include research on earthquake sources, attenuation of seismic waves, generation of building codes in areas subject to earthquakes, tomographic imaging of the earth's crust and deep interior, and identification and discrimination of nuclear explosions. The USNSN is an important step forward in providing data for the nation's seismological research. However, the USNSN, by itself, cannot fulfill many of the most important needs for seismic data across a broad spectrum of disciplines (e.g., public safety, public policy, critical facility siting, and basic science). An integrated National Seismic System (see Chapter 6), with dense regional networks in areas of high seismic potential, can fulfill these needs.

The unpredictability of the earthquake process makes it difficult to site dense instrument arrays. Limited resources preclude instrumenting the entire country with stations spaced at 50 km or less, which is the density in southern California. The high seismic potential of the major California fault systems provides a ready justification for dense networks there. But in the central and eastern United States, defining the seismic hazard to which a region may be subjected is a much more subtle and difficult problem, a fact that has contributed to the inadequate support for dense regional networks there.

The societal and scientific benefits that can accrue from dense regional networks are best shown by example. The following case studies document two earthquake sequences: the Whittier Narrows earthquake, which occurred within an existing regional network, and the Painesville, Ohio, earthquake, which occurred outside network coverage.

CASE STUDY:
THE 1987 WHITTIER NARROWS EARTHQUAKE IN THE LOS ANGELES METROPOLITAN AREA, CALIFORNIA

The moderate-sized (magnitude 5.9) Whittier Narrows earthquake occurred in the east Los Angeles metropolitan area at 7:42 a.m. (PDT) on October 1, 1987; it caused three direct fatalities and damage exceeding $350 million in many communities in Los Angeles and Orange counties.

The earthquake occurred in a densely populated area, but the location had seismological advantages: the focus was beneath the overlap of two regional networks, the 200-element Southern California Seismic Network jointly operated by the California Institute of Technology (CIT) and the U.S. Geological Survey (USGS), and the 24-element Los Angeles Basin Network operated by the University of Southern California (USC). Because of the excellent recordings of the earthquake afforded by the two existing regional networks, seismologists could quickly provide useful information to assist disaster response teams and emergency response officials. These data have also been the basis of detailed studies that have greatly improved understanding of the tectonics of the Los Angeles basin and of the seismic hazards facing the Los Angeles metropolitan area.

The Whittier Narrows earthquake occurred at 7:42 a.m. By 8:10 a.m., information on the earthquake's location, accurate to within 2 km, and on the magnitude of the main shock was available to emergency services personnel. (The CIT-USGS system has since been upgraded so that such data are available within 7-10 minutes.) This information was then used to help coordinate search and rescue operations. By 11:00 a.m., a focal mechanism was determined by using the data from the regional networks; it showed that the earthquake had occurred on a west-striking thrust fault—a subhorizontal fault along which the upper block had moved south, perpendicular to the strike of the fault. No such fault had been previously recognized in that area. Thus within a few hours after the earthquake, it was known that an earthquake with a magnitude of ~6 had occurred on a previously unrecognized thrust fault that could pose an additional earthquake hazard to the 12 million inhabitants of the Los Angeles metropolitan area (Hauksson, 1988, and Hauksson and Jones, 1989).

Data from portable arrays of seismometers installed in the epicentral region during the aftershock sequence were also used. However, because this aftershock sequence decayed particularly rapidly and the portable instruments were not installed until more than one day after the main shock, 90% of the magnitude >3.0 aftershocks occurred before data from the portable arrays were available. Thus, fundamentally important details of the Whittier Narrows aftershock sequence would have been irretrievable if data had not been obtained from the existing regional networks.

Using these data, Hauksson and Jones (1989) were able to construct a detailed, three-dimensional picture of the faulting during the Whittier Narrows earthquake sequence. The main shock and about half of its aftershocks occurred on the west-striking thrust fault; about one-third of the aftershocks, including the largest (with a magnitude of >5.3 on October 4), define a steeply dipping north-northwest striking fault with oblique right-lateral strike-slip movement. The near-vertical aftershock fault defines the edge of the subhorizontal main shock fault and may have confined the main shock slip.

This could explain why the patch of fault that slipped in the Whittier Narrows main shock was exceptionally small for an earthquake of this magnitude. The accommodation of the surrounding rock to the large strains produced by the main shock and largest aftershock was revealed in the small normal and thrust faults that were activated in the hanging wall above the main shock. The large quantity of data generated in the aftershock sequence was also used to develop a new, more accurate model of the seismic velocity structure of the Los Angeles basin.

Of course, the discovery of a previously unknown fault in the middle of the Los Angeles metropolitan area caused immediate concern. The most important unresolved question was the extent of the fault—was it limited to the Whittier Narrows area, or did it extend across the Los Angeles basin? Geologic investigations (Davis et al., 1989) had shown that the anticline that was the surface expression of the main shock fault at Whittier Narrows extended westward across the full width of the Los Angeles basin and seaward into Santa Monica Bay. Moreover, analysis of the shape of the anticline at Whittier Narrows and at another site near downtown Los Angeles strongly suggested that a thrust fault was buried beneath the anticline at those locations. The geologic information could not resolve whether or not the thrust fault extended under the full length of the anticline, nor could it show if the fault was currently active.

To answer these questions, Hauksson (1988) and Hauksson and Saldivar (1989) analyzed the data from small earthquakes recorded by the CIT-USGS and USC regional seismic networks. Because the networks had been in operation for many years prior to the earthquake, 15 years of pertinent data had already been archived. A search of these data for small earthquakes produced by thrust faulting showed that the full length of the anticline, from Whittier to Malibu, is indeed underlain by active thrust faults. These results have led to a reevaluation of the earthquake hazards facing the Los Angeles area. The new scientific findings are, in turn, being considered by local governments as they revise the seismic safety elements in their general plans. The data products of the permanent regional networks constitute an indispensable contribution to these important scientific and hazard assessment advances.

CASE STUDY:
THE 1986 PAINESVILLE EARTHQUAKE IN
NORTHEASTERN OHIO

The Whittier Narrows case study illustrates convincingly that high-quality information can be recovered when an event is "captured" by a dense seismic network. The Painesville, Ohio, case study illustrates the opposite situation—that important questions can remain unanswered when seismographic network coverage is absent.

On January 31, 1986, a modest-sized earthquake (magnitude 5.0) occurred in northeastern Ohio about 40 km east of Cleveland. It was felt in 11 states and in Ontario, Canada, and caused some minor damage (modified Mercalli scale intensity VI-VII) at distances up to 15 km from the epicenter. Its strike-slip mechanism implied a compressive east-northeast stress regime entirely consistent with previous events in the surrounding region. Thus, a first impression of this earthquake was that it was unremarkable, that is, very representative of the scattered, infrequent seismicity that characterizes much of the crust of the eastern United States.

Two facts, however, resulted in a greatly enhanced level of interest and concern about the Painesville earthquake. First, it occurred within 17 km of the Perry Nuclear Power Plant and produced high-frequency accelerations of 0.18 g there; second, it occurred within 15 km of the three deep fluid waste disposal wells that had injected over 1 billion liters of fluid into the earth's crust at depths of 1.8 km and at pressures exceeding 100 bars above ambient levels.

The subsequent detailed investigations of this earthquake by the U.S. Nuclear Regulatory Commission, the Ohio Environmental Protection Agency, the Cleveland Electric Illuminating Company, the U.S. Geological Survey, Weston Geophysical Corporation, and the Stauffer Chemical Company (operator of the two 1800-m Calhio wells) centered on the question of whether the fluid injection operations of the two deep waste disposal wells induced the Painesville earthquake. Several observations support a causal connection: (1) the Painesville earthquake was the largest event known to have occurred in the region; (2) in situ stress measurements in the Paleozoic sedimentary units overlying crystalline basement indicated the presence of high levels of deviatoric stress so that preexisting favorably oriented faults would be close to failure; (3) modeling (Nicholson et al., 1988) indicated that wellhead injection pressures of 110 bars could induce pore pressure changes from several bars up to 40 bars at 12 km from the wellbore, the actual value being quite sensitive to the confinement characteristics of the injection reservoir unit; and (4) pore pressure changes of this magnitude are known to have triggered earthquakes in other situations (e.g., Simpson, 1986).

Arguments favoring a natural rather than an induced origin for the Painesville event include the following: (1) the main shock's depth, although poorly constrained, places it in crystalline basement, not in the overlying Paleozoic rock where injection had occurred; (2) the main shock hypocenter was 12 km from the wellbores and approximately 3 km deeper than the injection depth; (3) northeastern Ohio had had a history of low to moderate earthquake activity prior to any injection operations; (4) the Painesville earthquake occurred 11 years after pumping had begun and did not correlate with any unusual pumping conditions; and (5) there was only one microaftershock, and no known prior seismicity in the crustal volume between the wellheads and the hypocenter.

These basic facts and all other data pertaining to the Painesville earthquake have been exhaustively examined by many investigators (principally Nicholson et al., 1988), but the unsatisfying conclusion is that a definitive decision on whether the main shock was natural or induced cannot possibly be made, given the quality and quantity of available seismological information.

Had a seismic network been in operation in the epicentral region, a more definite conclusion would likely have been reached. At the very least, two important additional pieces of evidence would have been available: tight constraints on the hypocenter's depth, and a much more sensitive test of whether microearthquakes had occurred around the wellbores during the 11-year pumping history. A focal depth confidently constrained to the 5-km centroid depth estimate, for example, would put the hypocenter about 3 km deep in crystalline basement. Without appeal to special fracture or joint pathways or pore fluid—especially if an absence of microearthquakes could be confidently established for the intervening crustal volume—a causal connection between the waste disposal wells and the main shock could be ruled out. An in-place seismic network could have provided the necessary information. Thus the Painesville event, by virtue of the lack of key data essential to the resolution of an important question, provides an excellent example of the value of operating seismic networks.

5

The U.S. National
Seismic Network

In previous chapters the panel has alluded to a sense of crisis concerning regional seismic networks. The crisis is engendered by two factors: (1) inadequate instrumentation that causes regional networks to fall behind the forefront of seismological research at an increasing rate, and (2) diversion of funds to support the planned U.S. National Seismic Network (USNSN). Most of the initial capital costs of the USNSN are being supported by funds previously used to support regional seismic networks in the central and eastern United States through an interagency agreement between the U.S. Geological Survey (USGS) and the U.S. Nuclear Regulatory Commission (see Appendix B). Therefore, in this chapter the panel briefly examines the plans for the USNSN and its dramatic impact on the future of observational seismology in the United States.

The USNSN is a new program being undertaken by the National Earthquake Information Center (NEIC) of the USGS with start-up funding from the Nuclear Regulatory Commission. The immediate objective of this cooperative effort is to establish a network of some 60 modern seismograph stations more or less evenly spaced throughout the United States east of the Rocky Mountains with satellite communications links to the NEIC. The ultimate goal of the effort is to record ground motion across a wide range of frequencies and with high dynamic range from all earthquakes nationwide above magnitude 2.5-3.0. This network fills an immediate need for uniform monitoring of earthquakes of magnitude 2.5-3.0 and above in the eastern United States. Near this magnitude level, earthquakes in populated areas are usually felt by more than a few persons. It is the responsibility of the

NEIC to provide a public statement on any earthquake felt in the United States. This network will allow NEIC to fulfill this function as well as to provide rapid reporting of damaging earthquakes and a high-quality data base for research on earthquake sources and the propagation of seismic waves.

The funding provided by the Nuclear Regulatory Commission carries the stipulation that it be used only for the purchase of equipment for the network east of the Rocky Mountains. At this time there are no funds for completion of the network west of the Great Plains and in Hawaii and Alaska. If completed nationwide, the USNSN will consist of approximately 150 seismic stations distributed across the lower 48 states, Alaska, Hawaii, Puerto Rico, and the Virgin Islands. This station density should be adequate to provide the capability to detect, locate, and quantify the energy release of earthquakes of magnitude 2.5-3.0 in all states except possibly Alaska. Such a capability will exceed that which exists today in many regions of the United States not currently monitored by regional networks. However, the USNSN will not, even if completed nationwide, eliminate the need for the existing regional seismic networks in areas of moderate and high seismicity. The principal purposes of such networks, described earlier in this report, are to detect earthquakes with very low magnitudes, down to around magnitude 1 in many cases, and to achieve highly accurate determinations of locations. Other important uses for which high station density is essential include earthquake hazard mitigation, earthquake prediction, estimation of strong ground shaking, and studies of the earth's crust and deep interior.

DESIGN OBJECTIVES

The USNSN is being designed to meet the following objectives:

• Detect and locate all earthquakes of magnitude 2.5-3.0 or greater within the United States;
• Report to the public all earthquakes of magnitude 2.5-3.0 or greater within the United States within 30 minutes;
• Minimize network development risk and cost;
• Minimize operational cost of the network;
• Locate the stations where the seismic "noise" is low;
• Measure the seismic signals over a wide range of frequencies and amplitudes; and
• Provide rapid distribution of the data products.

The capability of detecting and locating earthquakes of magnitude 2.5-3.0 or greater will ensure that most felt events are located with modest (±5-10 km) accuracy. The capability of reporting information on these earthquakes within 30 minutes is needed to allow the NEIC to issue rapid earthquake reports to emergency offices, government agencies, and the public. Regional seismic network centers provide similar information to the public and to emergency response centers in their areas.

The USGS objective is to have a reliable network yielding high-quality data within the six-year design and implementation period that began in 1988. The greatest possibility of incurring delays or cost overruns in projects such as the USNSN is generally associated with implementation of the central processing facility. For the USNSN, the USGS will minimize this risk by using a state-of-the-art seismic processing system recently developed for the NEIC. Only hardware that is currently available commercially will be used for the individual seismic stations. The most innovative feature of the USNSN, and therefore the one posing the most risk, is the use of satellite communications for the transmission of seismic data.

From past experience, the decision of whether to implement and operate seismic networks and arrays generally turns on projections of operational costs. To ensure the long-term stability of the USNSN, it is important that the annual operating cost of the network be kept low. The chances of the network surviving in a period of reduced funding are inversely proportional to the operating costs.

The NEIC plays a central role in distributing national and global seismic data to the scientific community. For data from the USNSN, the NEIC will establish procedures to ensure rapid distribution and equal access to the network data for all interested users. An example (in the lower 48 states) of a network configuration for the complete network of 150 stations is given in Figure 5. Only a few of these sites are fixed to date, but Figure 5 gives an idea of the station distribution and spacing that might be expected. The average station spacing is between 350 and 400 km, with a denser concentration of stations in the seismically active areas of the eastern and western United States.

STATION CHARACTERISTICS

Each USNSN station will be equipped with two sets of three-component seismometers, one set of conventional high-sensitivity instruments, and one set of instruments designed to respond linearly to strong ground motions experienced very near the epicenter of strong and moderate earthquakes. The system will provide 210 dB of dynamic range through 24-bit digitation at 80 samples per second (sps). The data will be recorded and transmitted in various bands. Not all of these bands will be recorded continuously; some will be "triggered" when the signal rises above a certain threshold. Each station will trigger independently as the signal conditions warrant. The various recording bands are characterized in Table 1. Individual stations will be supported by a microcomputer, a clock, a satellite transmitter and antenna, and solar panels and batteries for power.

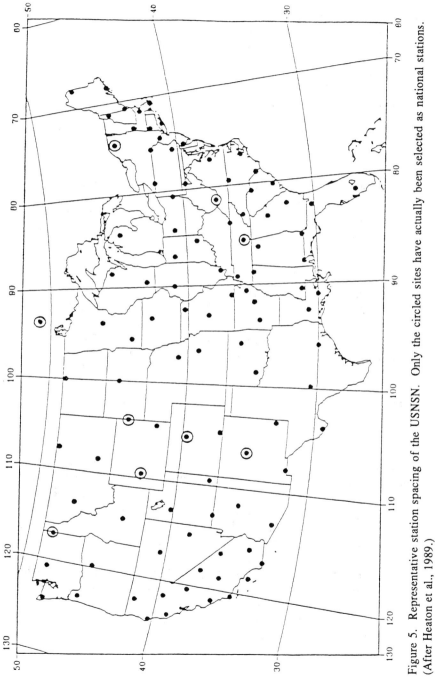

Figure 5. Representative station spacing of the USNSN. Only the circled sites have actually been selected as national stations. (After Heaton et al., 1989.)

Table 1. Recording Bandwidths for Seismic Stations

Bandwidth	No. Components	Frequency (samples per second)	Recording
Broadband	3-Component	40	Triggered
Strong motion	3-Component	80	Triggered
Long period	3-Component	1	Continuous
Short period	Vertical	13	Continuous

This station design should provide a wide range of seismic data useful for many purposes. In addition, the power and data transmission aspects of the design make the stations well suited for operations at sites remote from cultural activity, where seismic noise is likely to be low.

DATA TRANSMISSION

As mentioned above, data from individual stations will be sent to the NEIC via satellite. Each station will be equipped with a small satellite antenna less than 2 m in diameter. A master satellite receiving station will be located in the Denver area near the NEIC. The anticipated data capacity is a minimum of 2,400 bits/s per individual station (this can increase during peak periods) and 350,000 bits/s inbound and 50,000 bits/s outbound for the master stations at the NEIC. The data transmission protocol includes error detection, forward error correction, and packet retransmission.

DATA PROCESSING

A real-time seismic data processing system has recently been installed at the NEIC. This system is modular and will be expanded to meet the requirements of the national network through the use of additional hardware. Functions of the NEIC processing system include verifying and refining the triggered signal detections, determining the signal parameters, grouping or associating the signals from a single event, and determining a preliminary epicenter location. Other functions include maintaining an archive of waveform data with associated epicenter information, providing an interactive capability for a seismologist to review automated results, and producing final epicenter catalogs. In addition to epicenter catalogs, compact disks with read-only memory (CD-ROMs) containing all of the data collected by the USNSN will be produced and distributed routinely.

INTERFACE WITH NATIONAL AND REGIONAL STATIONS

The transmission of real-time seismic data by satellite from remote USNSN stations to the central processing facility at NEIC will be a major advance for U.S. observational seismology. Although satellite transmission of data is practiced on a limited scale by some Department of Energy and Department of Defense programs, the USNSN will be the first satellite-based instrument network available to the general seismological research community.

One of the major advantages of a satellite-based system over land-link telemetry is the flexibility of station siting: the dual limitations of line-of-sight links for radio transmission or availability of telephone line drop points are eliminated. The beneficial result is that station sites can be selected on the basis of low background noise or optimum station distribution rather than data transmission feasibility.

This new freedom in seismic network design and deployment raises a number of important issues concerning linking existing regional seismic network stations with the USNSN. Some of these are explored in an analysis by D.W. Simpson, which is reproduced in full as Appendix C to this report, and in the following chapter on the future of regional seismic networks.

6

The Future:
A National Seismic System

Forecasting the future of regional seismic networks is facilitated because that future is severly circumscribed. The panel has reported on the following persistent themes, already well set, which will control the next one to two decades of observational seismology in the United States:

• First, regional seismic networks, as currently configured and supported, do not have a long-term future; they will remain, at best, static in the western United States and will largely disappear in the East.

• Second, the rationale for development of the USNSN is compelling. However, since design and implementation of the USNSN are already well under way and funding for the eastern portion has already been secured, this is largely an after-the-fact finding.

• Third, the functions and data products of the USNSN are sufficiently different from those of the regional networks that the former cannot replace the latter. Even if completed nationwide, the USNSN will not eliminate the need for regional seismic networks.

The above themes, which the panel considers are amply supported in Heaton et al. (1989) and in this report, prompt reconsideration of the central recommendation of the Panel on National, Regional, and Local Seismograph Networks (Committee on Seismology, 1980), which is quoted in the preface. Implementation of a "rationalized and integrated" system consisting of a partnership between the USNSN and a confederation of existing regional seismic networks is also the central recommendation of the current Panel on Regional Networks. But now, 10 years later, the needs cited by the 1980 panel have become a crisis, the national network of "technologically advanced observatories" is close to becoming a reality, and a detailed functional framework and specific funding requirements have been identified. The total concept is called the National Seismic System.

ADVANTAGES OF A NATIONAL SEISMIC SYSTEM

It is important to reemphasize that the USNSN will not meet the need for data that can be obtained only through the dense spacing of individual stations in the typical regional network. The resolution required for the definition of local active tectonic structures cannot be achieved by the proposed national effort. Variations in propagation and seismic wave amplification, important in the assessment of earthquake hazards on a regional and local scale, cannot be measured by the USNSN. And finally, the USNSN cannot replace the training facilities and intellectual focus for seismological education and research that the regional networks currently provide at many universities throughout the country. However, the USNSN will provide a uniform, national earthquake recording capability that currently does not exist. Indeed, the planned national network and existing regional networks would complement each other, and together—if the former is developed and the latter continue to exist—provide an unprecedented source of seismological data for public services, education, and basic and applied research.

This combination of regional and national networks provides a unique opportunity to significantly advance seismic monitoring, data collection, data distribution, and seismological research in the United States within the next few years. This opportunity will be translated into reality only through close cooperation and coordination between the regional and national efforts and through the integration of certain aspects of their activities. The advantages that may be realized from a partnership of the regional and national network efforts include the following:

• Use of USNSN facilities could reduce communications costs. Expensive, often unreliable, and capacity-limited ground line communication links used by the regional networks are not very suitable for the transmission of seismic data. The satellite-based seismic data communications system being developed for the USNSN could revolutionize regional operations in that it will provide more reliable, more flexible, and less expensive communications service.

• Regional networks could provide maintenance and facility support for national network stations located within the monitoring area of the regional network. The national network would benefit through reduced operational costs. The host regional network would benefit by having direct access to the communications links of the USNSN.

- The sharing of communication links and other facilities would force both the regional and the national networks to adhere to certain standards of data quality and format.
- With such standards in effect, it would be easier to share and exchange software used in routine data analysis at both national and regional data centers. The standardization of data formats and software would allow data to be shared between regional networks and give easy access to the data from the national network.
- The national network could provide a framework or forum to draw the regional networks together to discuss and resolve common problems. The forum could prove to be an effective focus for the activities of the regional network operators and spur development of a body with a strong and unified voice on behalf of the concerns of the regional networks.

Thus, from both the state and the national perspective, there appears to be an opportunity for substantial benefit if the regional and national networks work together closely.

Finally, the panel examined the question of whether linkage with the USNSN is the only viable alternative for the regional networks and concluded that this is indeed the case. As has been shown, maintaining the status quo in network operations clearly is not an option. The most nearly related programs are the global network and portable array (PASSCAL) of the Incorporated Research Institutions in Seismology (IRIS). However, these programs are complementary to a National Seismic System, and IRIS has specifically avoided involvement with permanent arrays, although modernized regional networks would contribute greatly to such PASSCAL goals as three-dimensional imaging of the earth's crust. These reasons, combined with the fact that planning and funding for the USNSN are already well advanced, make a National Seismic System the best and only logical choice for the future of regional seismic networks. As the U.S. Geological Survey has already been assigned the role of developing the USNSN, it would play a major part in implementing the proposed system.

CURRENT AND PROJECTED COSTS OF A
NATIONAL SEISMIC SYSTEM

A National Seismic System cannot become a reality without the infusion of new funds. Currently, there are no new monies designated to (1) expand the USNSN to the western United States, (2) operate and maintain the USNSN beyond current NEIC resources, (3) replace the loss of $2 million in Nuclear Regulatory Commission regional network support, (4) replace and modernize aging and obsolete regional network instrumentation and equipment, or (5) provide for data links between the USNSN and the regional networks.

Based on its survey of network operators (Appendix A), the survey of the Ad Hoc Committee on Regional Networks (ACORN, 1986), and discussions with federal agency officials, the panel estimated that the FY 1989 annual apportionment of federal seismic network funds is approximately as follows: $1.5 million (USGS external networks), $3.0 million (USGS internal networks), $1.0 million (NEIC), $2.0 million (Nuclear Regulatory Commission), $2.0 million (DOE), and about $0.2 million (U.S. Bureau of Reclamation), yielding a total of nearly $10 million. This estimate does not include support for global networks or the restricted-use seismic operations of the Department of Defense. This, then, represents the approximate current expenditure for operations that would come under the aegis of the proposed National Seismic System.

The panel has not attempted a detailed analysis of projected costs for full implementation of a National Seismic System but has examined the question in sufficient depth to make firm recommendations in Chapter 7. For example, at least 65—and perhaps as many as 90—new stations will be required to complete the USNSN. At approximately $90,000 per station, the panel has conservatively estimated that $5 million will be needed for expanding the USNSN nationwide. Cost estimates for upgrading a typical regional network station range between $12,000 and $25,000, depending largely on whether broadband sensors are selected. If approximately one-third of the 1,500 regional network stations are modernized in the next five years, funding on the order of $10 million will be required for this element of a National Seismic System. These and other costs projected for full realization of a National Seismic System are included in Recommendation 6 of the next chapter.

7

Findings and Recommendations

1. It is important both for effective hazard mitigation and for scientific research that earthquakes within the United States be recorded regularly by a seismic system made up of networks operating on national and regional scales with long-term, stable financial support and uniform operating procedures. No such system now exists.

Recommendation. *The federal government should establish a more rational, coordinated, and stable means of support for the seismic networks of the United States either by consolidating funding and program management within a single agency or by assigning coordinative authority to a single agency for these purposes. Because of its assigned role in developing the U.S. National Seismic Network, it is recommended that this agency be the U.S. Geological Survey.*

2. Regional seismic networks with bases at universities or other research institutes, and operated in regions of moderate or high seismicity, play an essential and unique role in the recording and study of the nation's earthquakes. These networks and their central facilities provide a public service as local points for distribution of information on earthquake occurrences and on hazards posed by earthquakes. They provide data for basic and applied research on active tectonic structures within their particular regions and thus for prediction of possible earthquake activity, on the structural framework of the U.S. portion of the continent, and on other general seismological topics. They also provide realistic experience for the training and education of seismologists and other earth scientists.

Recommendation. The designer and operator of the proposed National Seismic System should consider that the services in research and education performed by regional seismic networks are necessary and integral components of that system.

3. The United States faces the imminent loss or technological obsolescence of its regional seismic networks. This is due to the lack of any government-wide policy for the long-term support of these networks and the restriction of funds within agencies that attempt to provide such support.

Recommendation. The federal government should provide long-term funding to stabilize the operation of regional seismic networks and, through a planned program of reasonable increases, to modernize these facilities.

4. The U.S. National Seismic Network now being developed in the eastern United States by the U.S. Geological Survey and the Nuclear Regulatory Commission embodies the correct approach to seismic monitoring in sections of the country where no regional networks now exist, to providing a standard base from which to report the occurrence of earthquakes, to providing data on earthquakes and seismic wave propagation characteristics on a continental scale, and to providing a framework for tying together the regional networks. Long-term support for the operations of the USNSN is needed, as is funding for extension of the USNSN to the western United States, Alaska, and Hawaii.

Recommendation. The U.S. Geological Survey should complete the USNSN in the eastern United States as designed and provide funding for its long-term operation and extension to the western United States.

5. A unique opportunity now exists to advance significantly, or at least to stabilize, earthquake monitoring, seismic data collection and dissemination, and, to some degree, seismological education and research in the next few years. This could be accomplished through the linking of the regional seismic networks to the USNSN in a National Seismic System.

Recommendation. The federal government should establish a National Seismic System through the technical linking and coordinated operation of regional seismic networks and an extended USNSN. This system should be supported by a single federal agency (probably the USGS because of its role in the USNSN), or one agency should be given authority for the coordination of its development and operation. Support for this system should be long term and should provide, through systematic planning, for modernization and for increases in operational costs due to inflation.

6. Currently, the federal government spends approximately $10 million per year to monitor and analyze the nation's earthquakes through seismic network operations. (This dollar amount is based on estimated funds budgeted for either internal or external seismic network support by the U.S. Geological Survey, the Nuclear Regulatory Commission, the Department of Energy, and the Bureau of Reclamation. The estimated amounts are for operational and basic analysis costs only; they do not include costs for special research using seismic network data.) Twenty percent, or $2 million, of the $10 million annual federal funding will be discontinued by 1992, when the Nuclear Regulatory Commission ends its program of supporting seismic networks in the central and eastern United States. Both the current budgetary levels and those projected for 1992 are seriously inadequate for carrying out a high-quality program of earthquake surveillance—either as now mandated by the National Earthquake Hazards Reduction Act of 1977 or as envisioned under future development of a National Seismic System.

Recommendation. The federal government should fund a National Seismic System at a level of $12 million per year, which is $2 million above current federal appropriations identified for seismic network operations. The $12 million base budget does not include funds necessary for regional network modernization. It will, however, ensure stabilization of existing network operations by providing some adjustment for the phased withdrawal of support by the Nuclear Regulatory Commission for seismic network operations in the central and eastern United States and some correction for the damaging effects of inflation caused by level federal funding during the past six years.

In addition to the annual operating costs cited above, not less than $15 million in new monies will be needed over a five-year period for (1) expanding the USNSN to full 50-state coverage, (2) funding satellite data links from the national center to the principal regional network operation centers, (3) upgrading computer facilities at the regional centers, and (4) standardizing and modernizing the regional network component of the proposed National Seismic System. This phased, one-time expenditure is considered necessary to fulfill the objectives of Recommendations 3, 4, and 5. Not less than $5 million will be required to complete the USNSN in the West and to complete satellite data links to key regional centers provided with upgraded computers. The creation of a National Seismic System presupposes modernization of at least a subset of the nation's 1,500 existing regional network stations. Conservatively, $10 million will be required to upgrade one-third of those stations to three-component, broadband sensing stations with fully digital data transmission.

Finally, as the National Seismic System is developed, it will be important to provide support for research in the universities that is based on the data produced by the regional seismic networks.

References

ACORN (1986). *Regional Seismic Networks: Scientific Achievements and Opportunities*, Ad Hoc Committee on Regional Networks, unpublished report, 8 pp.

Al-Shukri, H.J., and B.J. Mitchell (1988). Reduced seismic velocities in the source zone of New Madrid earthquakes, *Seismological Society of America Bulletin*, 78, 1491-1509.

Committee on Seismology (1980). *U.S. Earthquake Observatories: Recommendations for a New National Network*, Panel on National, Regional, and Local Seismograph Networks, Assembly on Mathematical and Physical Sciences, National Research Council, National Academy Press, Washington, D.C., 122 pp.

Committee on Seismology (1983). *Seismographic Networks: Problems and Outlook for the 1980s*, Workshop on Seismographic Networks, Commission on Physical Sciences, Mathematics, and Resources, National Research Council, National Academy Press, Washington, D.C., 62 pp.

Davis, T.L., J. Namson, and R.F. Yerkes (1989). A cross section of the Los Angeles area: seismically active fold and thrust belt, the 1987 Whittier Narrows earthquake, and earthquake hazard, *Journal of Geophysical Research*, 94, 9644-9664.

FEMA (1988). *National Earthquake Hazards Reduction Program: Fiscal Year 1987 Activities*, Federal Emergency Management Agency, report to the U.S. Congress, 183 pp.

Goltz, J. (1985). *The Parkfield and San Diego Earthquake Predictions: A Chronology*, special report by the Southern California Earthquake Preparedness Project, Los Angeles, Calif., 23 pp.

Hauksson, E. (1988). Thrust faulting and earthquake potential in the greater Los Angeles basin, southern California, *EOS, Transactions of the American Geophysical Union*, 69, 1305.

Hauksson, E., and L.M. Jones (1989). The 1987 Whittier Narrows earthquake sequence in Los Angeles, southern California: seismological and tectonic analysis, *Journal of Geophysical Research*, 94, 9569-9589.

Hauksson, E., and G.M. Saldivar (1989). Seismicity and active compressional tectonics in Santa Monica Bay, southern California, *Journal of Geophysical Research*, 94, 9591-9606.

Hearn, T.M., and R.W. Clayton (1986). Lateral velocity variations in southern California, Parts I and II, *Seismological Society of America Bulletin*, 76, 495-520.

Heaton, T.H. (1985). A model for a seismic computerized alert network, *Science*, 228, 987-990.

Heaton, T.H., D. Anderson, W. Arabasz, R. Buland, W. Ellsworth, S. Hartzell, T. Lay, and P. Spudich (1989). National Seismic System Science Plan, *U.S. Geological Survey Circular 1031*, 42 pp.

Hill, D.P. (1987). Seismotectonics, *Reviews of Geophysics*, 25(6), 1139-1148.

Humphreys, E., R.W. Clayton, and B.H. Hager (1984). A tomographic image of the mantle beneath southern California, *Geophysical Research Letters*, 11, 625-627.

Lander, J.F., and P.A. Lockridge (1989). *United States Tsunamis (including U.S. Possessions), 1690-1988*, Publication 41-2, National Geophysical Data Center, National Oceanic and Atmospheric Administration, Boulder, Colo.

Lay, T. (1987). Structure of the earth: mantle and core, *Reviews of Geophysics*, 25(6), 1161-1167.

Nicholson, C., E. Roeloffs, and R.L. Wesson (1988). The northeastern Ohio earthquake of 31 January 1986: was it induced? *Seismological Society of America Bulletin*, 78, 188-217.

Simpson, D.W. (1986). Triggered earthquakes, *Annual Reviews of Earth and Planetary Sciences*, 14, 21-42.

Simpson, D.W., and W.L. Ellsworth, Convenors (1985). Proceedings of a Symposium and workshop, regional seismographic networks, past-present-future, October 1985, Knoxville, Tenn., *U.S. Geological Survey Open-File Report*.

APPENDIXES

Appendix A

Survey of Regional Seismic Networks

In preparing this report, the panel considered it essential that the current status of regional seismic networks be examined. A rapid assessment of basic information was performed through a mailed questionnaire and follow-up telephone call. The panel believes that all major regional networks were contacted and that any omissions would have only a minor effect on the tabulation. The questionnaire, Figure A1, was sent to all network operators.

The results of the survey have been divided into two parts. Those relevant to budgetary considerations are discussed in detail in Chapter 3 of the main report, (in the section titled, "Are Regional Networks Cost-Effective?"). Other survey results are summarized in Table A1, and several findings that characterize the overall activity of regional seismic networks in the United States are highlighted below:

• There are nearly 50 operators of regional networks in the United States. A rough breakdown with some overlap is as follows: at least 24 universities operate regional networks (some combining several different networks into one overall operation); about 8 federal agencies operate some 14 networks; at least 6 networks are operated by state agencies; and several networks are operated by private utilities or geotechnical firms.

• A total of 1,508 seismograph stations are operated in permanent or quasi-permanent, regional or quasi-regional networks. (This total is probably accurate to within 5%.) Eight California or California-Nevada networks account for over 600 stations, or about 40% of the total.

• The U.S. Nuclear Regulatory Commission currently provides full or

Table Al. Regional Seismic Networks of the United States (circa 1989): A Summary

Institution	Area	Stations[1]	3-Comp	Digital[2]	Upgrade[3]	Computer[4]	Student/FTE	Support[5]
1. Univ. Washington	WA,OR	123	3	0	1	1	8/6	USGS,DOE
2. Lawrence Livermore National Lab.	CA,NV,UT	4	4	4	4	1	0/0	DOE
3. Boise State Col.	ID	3	0	0	0	0	4/0.4	State
4. Univ. Idaho	ID	4	0	0	0	1	4/1	State
5. INEL	ID	6	1	0	1	0	0/0	DOE
6. Ricks College	ID	5	0	0	0	0	3/1	Private
7. Mont. Bur. Mines	MN	12	0	0	0	—	2/0.5	State
8. BuRec	WY	16		0	0	0	3/0	BuRec
9. USGS Golden	NV	55	2	0	0	8	3/0.5	DOE
10. Univ. Utah	UT,ID,WY	82	4	0	2	8	4/2.5	USGS,BuRec,State
11. Bureau of Reclam.	CO	15	2	0	5	3	0/0	BuRec
12. Bureau of Reclam.	CO	7	0	0	7	3	0/0	BuRec
13. N. Ariz. Univ.	AZ	7	0	0	7	2	3/1	State,Private
14. Los Alamos	NM	8	3	0	3	5	0/0	DOE
15. N. Mex. Tech.	NM	15	0	0	0	2	6/3	DOE,State
16. Southern Calif.	CA	24	12	1	12	2	2/1	USGS/City
17. Univ. Calif. San Diego	CA	10	10	10	0	8	3/3	NEIC,State,Private
18. Univ. Calif. Berkeley	CA	20	6	4	4	4	10/3	NEIC,State,Private
19. Univ. Nevada	NV,CA	62	10	8	29	1	7/4	USGS,State
20. CDMG	CA	8	1	0	8	4	2/0.5	State
21. USGS Menlo Park	CA	327	12	0	0	1	0/0	USGS
22. Calif. Water Res.	CA	18	4	0	7	1	2/1	State
23. USGS Hawaii	HI	52	14	0	0	3	2/1.25	USGS
24. USGS/Caltech	CA	200	15	2	5	1	3/1	USGS,State,Private
25. Okla. Geol. Survey	OK	11	1	1	4	0	0/5	NRC,DOE,State
26. Univ. Mich	OH,IN,MI	14	0	0	4	10	2/0	NRC

27.	St. Louis Univ.	MO,AR,TN,IL	42	2	1	1	10	6/8	USGS,NRC,Private
28.	Kansas Geol. Survey	KS,NB	15	0	0	0	6	0/1	NRC
29.	Univ. Puerto Rico	PR	21	0	0	0	5	5/1	Univ.
30.	VPISU	VA	16	4	0	5	9	0/0	NRC,State,Private
31.	Del. Geol. Survey	DE	5	0	0	1	2	0/0	State
32.	South Carolina	SC	6	0	0	0	0	3/0	Private
33.	Univ. Kentucky	KY	8	0	1	6	1	2/0	State
34.	Tenn. Valley Auth.	TN	21	2	2	0	—	0/0	TVA
35.	Ga. Tech.	GA,AL	21	4	0	3	2	7/2	NRC,State,Private
36.	USGS South Carolina	SC	19	5	0	10	12	1/0	NRC
37.	Duke Power	SC	4	0	0	0	—	0/0	Private
38.	Savannah River	SC,GA	4	1	0	3	1	0/0	DOE
39.	Memphis State/CERI	AR,TN,GA,NC	30	4	1	4	7	9/3	NRC
40.	Woodward/Clyde	NY,NJ	38	3	5	11	2	0/0	Private
41.	Lamont-Doherty	NY,N.England	32	3	1	0	10	2/0.5	NRC/USGS
42.	Boston Col./Weston	N.England	29	1	14	0	0.5	6/3	NRC
43.	Mass. Inst. Tech.	N.England	9	2	0	0	1	2/0.4	NRC
44.	Univ. Alaska	AK	45	2	0	4	2	4/1	USGS
45.	Univ. Colorado	Aleutians	14	1	0	0	10	3/1	USGS
46.	Lamont-Doherty	Aleutians	18	6	1	1	2	0/0	DOE,USGS
47.	SUNY Stony Brook[6]	NY	3	0	0	0	—	—	NRC
	Totals:		1,508	144 (9.7%)	40 (2.7%)	153 (10.1%)		123/55.8	

[1] Number of stations actually operated; often more are recorded.
[2] Completely digital; sensor to recorder.
[3] Number of stations significantly upgraded in last 5 years.
[4] Age, in years, of computer in use for network recording.
[5] "Private" almost always indicates utility company support.
[6] Network scheduled to close March 1989.

Panel on Regional Networks
1988 Information Survey

A MEMBER OF THE PANEL WILL CONTACT YOU SOON BY TELEPHONE REQUESTING THIS INFORMATION. AT THAT TIME YOU WILL ALSO BE ASKED TO FORWARD A LISTING OF STATION COORDINATES.

1. Network name and operating institution _____

_____

2. Total number of stations _____

3. Total current annual funding for network operations alone, exclusive of research $_____

4. Relative contributions to current network support:

    a. Federal
        USGS   _____%
        NRC   _____%
        Other   _____%
        Total Federal   _____%
    b. State   _____%
    c. Private   _____%
        TOTAL   __100__%

5. Number of three-component stations in network   _____

6. Number of completely digital stations in network
   (i.e., from sensor through recording, not just
   recording alone)   _____

7. Number of stations significantly upgraded during last
   5 years   _____

8. Age of computer in use for network recording   _____

9. Number of students (undergraduate and graduate)
   involved in network seismology doing either analysis
   or research on network data. Please estimate the total
   FTEs (full-time equivalents) also.   _____Students

                                                       _____FTEs

Figure A1. Sample questionnaire.

partial support for some 14 seismic networks in the eastern United States, totaling some 241 stations. This represents only 15% of the national total but more than 70% of the regional network seismic stations operated east of the Rocky Mountains. Complete phase-out of this Nuclear Regulatory Commission support is scheduled for 1992.

• Less than 10% of the seismic stations currently operated by regional networks record the full three-dimensional seismic wavefield. Less than 4% are fully digital from sensor to recorder. Only 10% of the stations have been significantly upgraded in the last five years.

• Some 123 students in seismology were fully or partially supported by seismic network operations in 1988.

Appendix B
Interagency Agreement Between the United States Nuclear Regulatory Commission and the United States Geological Survey

PURPOSE: The purpose of this Interagency Agreement is to set forth a plan for establishing a network of seismic stations for monitoring seismicity in the Eastern and Central United States agreed to by the United States Geological Survey (USGS) and the United States Nuclear Regulatory Commission (NRC).

BACKGROUND: Frequency of occurrence, geographical distribution, and magnitude of earthquakes are important characteristics in assessing the seismic hazard of a region and establishing the design and construction criteria for a critical facility at a specific site. These characteristics are known collectively as the seismicity of a region and can only be determined through the operation of networks of seismometers that record earthquakes and analysis of these recordings.

Under the Earthquake Hazards Reduction Act of 1977 (Public Law 95-124) the USGS is charged with assessing the earthquake hazard and developing earthquake prediction systems in those areas of the United States subject to moderate-to-high seismic risk. The goal of the USGS program is to mitigate earthquake losses that can occur in many parts of the United States by providing research, evaluations, and earth science data for land-use planning, engineering design, and emergency preparedness decisions. Specific objectives of the USGS program are: (a) to evaluate the earthquake potential of the seismically active areas of the United States; (b) to provide assessments of earthquake potential of the seismically active areas of the United States; (c) to provide assessments of earthquake hazard and risk in developed regions exposed to the earthquake threat; (d) to predict damaging earthquakes; (e) to provide data and information on earthquake occurrences to the public and scientific community; and (f) to provide data and estimates of the level and character of strong earthquake shaking to be used in earthquake-resistant design and construction. To carry out this work the USGS supports in-house research in geology, geophysics, and engineering as well as significant supporting activities. This program is augmented and strengthened through support of complementary scientific investigations at universities, state agencies, and private companies. USGS earthquake hazards activities are coordinated with related efforts in the Federal Emergency Management Agency, the National Science Foundation, and the National Bureau of Standards through the National Earthquake Hazards Reduction Program.

The NRC has certain responsibilities for ensuring public health and safety in regard to potential hazards associated with nuclear power plants, radioactive waste disposal facilities, and other activities involving radioactivity. Thus, the NRC has a strong interest in the delineation, assessment, and mitigation of earthquake hazards in the United States, particularly as they pertain to nuclear power plant and radioactive waste disposal facility siting, design, construction, and operation. Because most of the nation's nuclear power plants are located east of the Rocky Mountains, the NRC has provided special support for earthquake hazard delineation in the central and eastern regions of the United States. These NRC efforts contribute to the goals of the National Earthquake Hazards Reduction Program as well as the NRC's more immediate needs. NRC-supported studies contribute to (a) the better definition of seismicity by determining the location, magnitudes, recurrence rates, and special characteristics of earthquakes; (b) the quantification for seismic hazard and the reliability of seismic hazard assessments; and (c) the definition of the relationships between seismicity of a region and its geologic structure and tectonics.

Given that the objectives of the USGS and the NRC regarding regional seismicity are so interrelated, they wish to pool their resources to establish a modern seismographic network in the United States east of the Rocky Mountains.

OBJECTIVE: The objective of this agreement is to establish a network of modern seismographic stations for monitoring the seismicity in the United States east of the Rocky Mountains.

This objective implies a significant change in approach to monitoring the seismicity of this part of the United States and the eventual replacement of NRC's existing regional seismographic networks with an integrated network of seismographic stations covering the entire United States east of the Rocky Mountains. The general strategy for the new network is outlined in a 1980 report, *U.S. Earthquake Observatories: Recommendations for a New National Network,* by the Panel on National, Regional, and Local Seismographic Networks of the National Research Council.

ELEMENTS OF AGREEMENT:

1. Beginning with Fiscal Year 1993, the USGS will assume full responsibility for monitoring earthquakes in the United States east of the Rocky Mountains. This monitoring will be accomplished through a new integrated network of state-of-the-art seismographic stations.

2. A joint USGS/NRC working group shall prepare recommendations by November 30, 1986, for a plan for the development, testing, installation, and operation of the new seismographic stations. Based on these recommendations, the NRC and the USGS will develop an amendment to this agreement that will set forth the plan for the development, testing, installation, and operation of the new stations. The plan will include:

a. The number and location of the stations to be built.

b. A budget and schedule for acquisition of the network hardware and for the commissioning of stations.

c. A protocol for timely access to times series and parameter data recorded by the new network. The protocol will encompass access to data by federal agencies, cooperating/operating institutions, and the general public.

d. A protocol describing the initial and the long-term working relationship among the NRC, USGS, and cooperating/operating institutions.

3. After the plan has been agreed to by the USGS and the NRC, the NRC will provide to the USGS a total sum of $5 million on the following schedule subject to the availability of appropriations:

Fiscal Year	Amount
1987	$500K
1988 through 1992	(The schedule of payments for this period will be set by amendment of this agreement following the completion of the recommendations of the joint working group.)

These funds will be used exclusively to acquire the permanent equipment, including operating software, necessary to establish the new network.

4. The USGS shall assume full responsibility for the continuing operation of the new stations as soon as reasonable after they are installed.

5. Progress shall be jointly reviewed by the NRC and the USGS in semi-

annual meetings. Minutes of the meeting will be taken and provided to the cognizant NRC and USGS management. Any unresolved issues will be highlighted as appropriate.

6. By entering into this agreement, the USGS does not assume responsibility for any existing seismic monitoring equipment or other related activities currently supported by the NRC through contracts or other legal instruments.

7. Either party to this agreement may terminate the agreement by providing 90 days' written notification.

U.S. Nuclear Regulatory Commission

BY: _____

NAME:

TITLE: Executive Director for Operations
         USNRC

DATE:

U.S. Geological Survey

BY: _____

NAME: Dallas Peck

TITLE: Director, USGS

DATE:

Appendix C
A Revitalization of
Regional Seismic Networks:
Implementation Strategies

David W. Simpson
Lamont-Doherty Geological Observatory
Palisades, NY 10964
September 1988

INTRODUCTION

The decision of the U.S. Nuclear Regulatory Commission to phase out its support of regional networks in the eastern United States, and to support the establishment of a National Seismic Network by the USGS, has led to both a short-term crisis for the support of regional earthquake studies in the East and a long-term opportunity for revitalization of regional network data gathering. It is not the purpose here to make the case for the breadth of scientific or practical opportunities provided by regional networks. Various documents from IRIS, the National Academy of Sciences and the USGS have described in detail the range of new opportunities that can be explored with data from modern seismographic instrumentation. A group led by Tom Heaton is developing a Science Plan for a National Seismic System, which concentrates on the specific contributions that can be made by networks on a regional scale.

From discussions in various forums over the past two years, the following have emerged as some of the main areas of consensus on the developments necessary to improve the state of regional seismology in the United States:

• The U.S. National Seismic Network (USNSN) provides, both in concept and implementation, a model that has wide ranging implications for regional seismic studies. Broadband waveform data from the network will be used in the analysis of regional earthquakes. The satellite communication system that forms the backbone of the USNSN will have capacity to handle regional network data as well.

• Even with the establishment of an enhanced National Seismic Network, there will continue to be an essential role for regional-scale networks in monitoring low-magnitude seismicity and defining earthquake patterns with a scale and resolution that are geologically significant.

• If regional seismology is to survive, both financially and intellectually, major changes are required. The financial future depends on decreasing operating costs and establishing new funding sources, especially at the state level, for stable support of operation. The intellectual future depends on making major changes in the way regional networks are operated in order to provide data that are sufficient to meet the challenges of new interests and techniques in analysis of the complete seismogram.

• There continues to be a wealth of opportunity for research in seismology at the regional scale. Fundamental problems in geotectonics and earthquake prediction remain to be solved. There is increasing interest in monitoring nuclear test ban treaties with seismic stations at regional distances.

• There are no technical limitations to acquiring data that satisfy current and anticipated research needs. Broadband sensors with digital conditioning and telemetry are now capable of reproducing earthquake motions over the complete amplitude and spectral range of interest in regional seismology.

• A reassessment of the mode in which regional network data are processed, archived, and distributed should be carried out in concert with any major changes in field equipment for data collection.

• A funding strategy should be developed that provides for the capitalization of new equipment for regional networks, stabilizes the long-term support for routine network operation (maintenance, processing, and catalog generation) and encourages the growth of funding for research based on data from an improved and integrated national/regional network system.

REGIONAL NETWORKS—CAPABILITIES AND LIMITATIONS

Regional telemetered networks, of the type now in operation in many parts of the United States, were originally installed with the primary purpose of locating large numbers of small-magnitude events as a mapping tool in defining active faults and in determining modes of deformation based on fault plane solutions. As an increasingly detailed picture of seismicity along major active zones emerged, it has become obvious that regional seismic networks are also important tools in monitoring intermediate-term regional strain (through changes in seismicity). Deviations from the stable background seismicity, both in space and time, are extremely important in unraveling the details of regional tectonic processes and may play an essential role (through the detection of foreshocks and other short-term changes in seismicity) in earthquake prediction. Both the magnitude range ($m \ll 3$) and spatial resolution necessary to observe these processes demand monitoring on a regional scale with a station density that is beyond the scope of a national network.

The use of networks telemetering relatively low-dynamic-range, narrow-bandwidth data to a central recording facility has served well the purpose of studying the spatial and temporal distribution of seismicity. The low cost per station has allowed for a relatively large number of sites. The high magnification at high frequencies and recording at a central station with common time base have allowed for accurate timing of body phases. Both of these factors—a dense network and common timing—are necessary to provide accurate location and a sensitive detection threshold in seismicity studies. There have been trade-offs, however. Inherent in the low cost and the analog telemetry available in the 1960s, when these systems were originally designed, are severe restrictions on dynamic range (less than 50 dB) and bandwidth (1-20 Hz). The major factor limiting the quality of data recorded is the continuous analog telemetry, often by telephone. With rising telemetry costs, the maintenance of this weakest link in the system has also become one of the major costs in operating the networks. At the same time, techniques for the analysis of seismic data have developed to the level where the quality of the data obtained now lags behind the expectations of those wishing to use them. Broader bandwidth and higher dynamic range are required for a wide variety of new studies in earthquake source mechanics, seismic wave propagation, and structure of the earth.

THE U.S. NATIONAL SEISMIC NETWORK

The USNSN was designed to locate earthquakes above magnitude 2.5 throughout the United States to serve the monitoring purposes of the National Earthquake Information Center (NEIC) and the Nuclear Regulatory Commission (NRC). As such its primary goals are to provide relatively uniform coverage throughout the United States and rapid data telemetry to a central facility in Golden for near-real-time hypocenter determination.

As a monitor of seismicity at the national level, the USNSN will provide for the consistent and stable location of earthquakes above magnitude 2.5, both serving the immediate reporting purpose of the NEIC and providing the foundation for eventually producing a long-term stable catalog of U.S. earthquakes, which will have broad application in studies of seismicity and hazard assessment.

As a facility for collecting and distributing waveform data from a continent-scale array of first-order broadband stations, the USNSN will make it possible to carry out studies of earthquake sources, wave propagation at regional distances and the structure of the continental lithosphere and earth's deep interior.

As a national communication network for seismological data, the USNSN holds perhaps the greatest potential for improving regional earthquake studies. The USNSN satellite telemetry system can provide a low-cost means of collecting and distributing a significant component of regional network data. The existence of a national system for data distribution will also help to encourage the interchange of data among various networks.

Details of the instrumentation and organization of the USNSN are described in various documents from the Branch of Global Seismology. For the purposes of the discussion here, the following points are highlighted:

• The USNSN instrumentation provides data adequate for full waveform analysis of significant regional, national and global earthquakes (i.e., on-scale recording of three-component ground motion in the bandwidth 100 s-30 Hz).

• The density of stations within the USNSN (relatively uniform spacing of 110 stations over the continental United States) is a reasonable compromise between cost and quality for monitoring on a national scale.

• The USNSN, through the NEIC, will have the facility for rapid location of significant earthquakes.

• The USGS will be responsible for the archiving and timely distribution of all USNSN data.

• The USNSN will have the facility to provide near-real-time access to waveform data of interest in regional earthquake studies.

• The USNSN telemetry system will have a capacity sufficient to handle significant amounts of regional network data.

NATIONAL-REGIONAL NETWORK INTERACTIONS

The USNSN will provide data for a variety of new approaches to studying earthquake sources and seismic wave propagation that will have direct application to problems on a regional scale. While it may meet some of the current requirements placed on regional networks, it will not replace the key characteristic of regional networks, namely the close station spacing and thus the capability for high spatial resolution of earthquake hypocenters. The USNSN must be seen, therefore, as a complement to the current activities of regional networks and not a replacement.

Centers for regional seismic studies, with regional networks as an important component, will continue to play an important role in providing a complement to a national program to:

• provide denser coverage to focus on areas of special interest (higher seismicity, significant tectonic problems, higher seismic risk, critical engineering structures);

• act as a regional center for coordinating response to earthquakes and interaction with the public on questions of regional seismicity;

• provide a regional focus for seismicity studies and broad-based seismological research; and

• provide for the cataloging, archiving, and distribution of data on a regional basis.

In terms of network operation, the primary goals of an integration between the USNSN and regional networks should be to:

• remove the spectral barriers inherent in the current collection of earthquake data, allowing data with appropriate bandwidth to be applied to regional, national, and global earthquake studies;

• make use of modern digital signal conditioning, processing, and telemetry to decrease the cost of data collection and at the same time improve data quality; and

• develop a system of data collection and distribution that allows near-real-time access to data at regional centers and coordinate the access to all levels of network data by all interested users.

The design route that the USGS has taken has involved a number of decisions that result in a system that has potential for development well beyond the original monitoring purpose of the network. This applies not only in the inherent capabilities of the network itself, but especially in the area of interaction of the USNSN with regional networks. The network design throughout involves digital signal conditioning and telemetry of broadband three-component ground motion, providing high-quality data suitable for many research applications on regional, national, and teleseismic earthquakes. The satellite telemetry systems provide a vehicle for national communication of seismic data, which has the capacity to extend well beyond the demands of the USNSN itself.

Beyond the complementary roles that they can play in data collection, the importance of regional and national networks in stimulating and focusing research on problems of different scale is also important. While arguments of scale and efficiency might be used to make the case for centralized control of a complete national recording system, the reality is that local participation in data gathering is essential in the stimulation of research on regional problems. A regional focus in seismicity studies is also important for increasing public awareness of earthquake problems and in interactions with the public and news media following felt earthquakes.

DATA REQUIREMENTS FOR REGIONAL NETWORKS

The fundamental requirements for regional networks are the same as those for a national network.

As a monitoring tool: near-real-time access to arrival times and amplitudes from all stations

As a research tool: full waveform recording of all events of interest

The primary difference is in the number of stations required; regional studies require a station density that is at least an order of magnitude higher than for national monitoring. To instrument regional networks to the required density entirely with stations of USNSN quality is financially impractical. Thus, there is the need to develop a less expensive means of providing the special data needed to satisfy the additional requirements of regional networks.

Some compromises are required. Regional studies concentrate on data from earthquakes at relatively short distances and often from source zones that are known a priori. Therefore, the frequency range of interest is higher and narrower than for a national network and the distribution of stations need not be uniform, but can concentrate on regions of known seismicity. In most networks, sufficient experience has been gained with the character of local seismograms to allow for considerable automation in the identification of events and extraction of parameters (arrival times, amplitudes etc.). Thus, at the lowest level, it may be possible to develop one class of regional stations which provide only limited parameter data and short waveform segments from events.

One model for how a national and regional network might interact is shown in Figure C1. Types of data and telemetry links are summarized in Figure C2 and Table C1. Within a given region, the national network produces data from a relatively small number of broadband first-order stations, and the national network satellite link provides real-time telemetry to the national center and back to the regional center. In those areas where station density and communication links make it feasible, regional nodes (in some cases co-located with a national network station) gather data from a dense cluster and use the same USNSN satellite telemetry link back to the regional center. Regional network nodes would be capable of automatic event detection, parameter determination, hypocenter location, coordination of communication protocol, and backup recording. These nodes might be located at cooperating institutions (e.g., a local college), providing support for maintenance and local recording.

For most networks, satellite telemetry might provide the only data continuously received at the regional network center. This data stream would consist of:

continuous long-period data,	from each USNSN station within
broadband events,	the region
continuous short-period monitor	

plus

short-period events,	from all nodes in areas of high
derived parameters,	station density
selected continuous monitors	

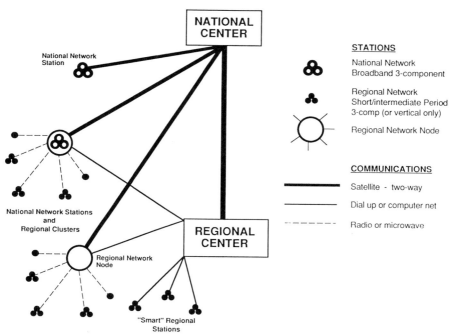

STATIONS

National Network
Broadband 3-component

Regional Network
Short/intermediate Period
3-comp (or vertical only)

Regional Network Node

COMMUNICATIONS

Satellite - two-way

Dial up or computer net

Radio or microwave

Figure C1. An integrated national/regional network system. National network satellite telemetry link provides telemetry of all event data and some continuous short-period data from regional clusters to regional centers in near real time. Regional centers have access to data from all stations via dial-up or computer network.

These data sources would form the backbone of the regional monitoring system. Additional "smart" regional stations would be located where required to provide the necessary station density. Since the backbone network is intended to provide the basic monitoring, these additional remote stations need not have continuous telemetry. They would be provided with sufficient intelligence to detect and store event parameters and waveforms, to be regularly or automatically accessed, decreasing communication costs.

After implementation of the USNSN, the major components of the system that need development are the hardware for regional nodes and smart stations (both processors and communication) and the software to control the data flow. Many of the concepts for the smart station and regional node hardware are in various stages of development by regional network groups. Sufficient experience should now be available to set specific guidelines for the development of both node and smart station processors that would be acceptable by most networks. An immediate task for regional network operators should be to initiate a concrete plan for development of these components.

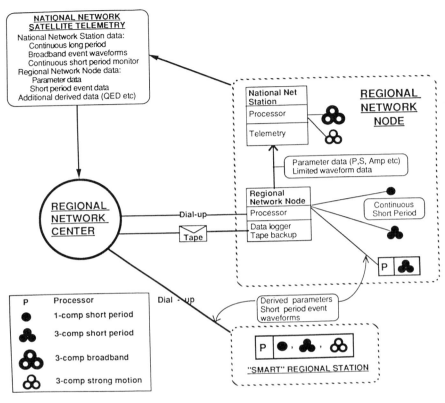

Figure C2. Data sources and telemetry for regional seismic networks.

RECOMMENDATIONS

USNSN Deployment

In those parts of the country where regional networks now exist, the development of an integrated national/regional network system will depend on close coordination between regional network operators and the USGS.

- Continue development of a plan for siting USNSN stations.
- Take into consideration possible sites for regional network nodes in choosing USNSN station locations.
- Involve regional network operators in site selection, installation and maintenance of USNSN stations.

Table C1. National Seismic Sytem: Data Sources

US National Seismic Network

100+ stations at 100-300 km spacing

Long Period	1	sps	0.3 hz	continuous	⎫
High Frequency	80		30	strong events	⎪
Broadband	40		15	events	⎬ Satellite telemetry
Short Period Monitor	13.3		5	continuous	⎪
QED				event parameter data	⎭

Regional Clusters

10's of stations per cluster at ~10 km spacing
Multiple clusters per network

Short Period Events	100+	sps	50 hz	events	⎫
Short Period Monitor	13.3		5	continuous	⎬ Satellite, dial-up &/or tape
Parameter Data					⎭
Short Period Monitor				continuous	Local helicorder

"Smart" Regional Stations

Remote dial-up stations
As required to fill gaps in regional nets

Short Period Events	100+	sps	50+ hz	events	dial-up, packet radio
Parameter Data					

USNSN satellite telemetry — continuous near-real-time to Regional Centers
— possible broadcast mode

Eventual goal — All event waveforms recorded from three component stations
— Digital signal conditioning from seismometer to recorder
— Two-way communication and automatic polling of remote stations

Parameter Data — Single station - phase times, ampliudes, duration, first motion etc
Clusters — -ditto- plus preliminary event location
State of health, system parameters

Implementation Experiment

To a large extent the use of the USNSN telemetry for regional network data is more an experiment in data communication than seismology.

• Accelerate the initial deployment of the USNSN by providing satellite telemetry to universities or regional networks with existing broadband stations to start experimentation with telemetry and data collection.

• Start with those networks that have regional node configurations to experiment with the concept of national/regional network integration.

• Develop specific protocols for interaction between the USNSN, regional networks, and other university groups.

Hardware Development

New station and processor hardware for regional networks will be required to see the development of a completely integrated National Seismic System.

• Network operators and engineers should meet to draw up specifications of additional hardware components required to complete the integration of a national/regional network.
• Cost and integration with existing new development programs (USNSN and IRIS) should be major design factors.
• One or two groups should be identified to lead a national project in the development of these systems.
• Special funding should be found to support this development.

Software Development

The transition to a new system of data collection provides the opportunity to carefully reevaluate the ways data are processed, archived, and distributed. Considerable standardization in computers for data processing already exists between regional networks (primarily Unix). Data collection and distribution through the USNSN satellite link will impose one level of standardization in data formats and communication.

• Serious consideration should be given to the advantages of developing standard software for the initial processing and cataloging of regional network data (i.e., below the "research" level) to simplify data exchange and improve quality control.
• One or two groups should be identified to lead a national project in the development of new software.
• Special funding should be found to support this development.

Funding Strategy

A new National Seismic System, combining a national network with regional programs, will require capitalization beyond that available from existing programs.

• A Science Plan should continue to be developed that clearly identifies the unique contributions of regional networks and is aimed at those agencies that can benefit from regional studies.
• The Committee on Seismology's Panel on Regional Networks should prepare a realistic profile of the funding required to establish an integrated National Seismic System of national and regional networks.

• The funding profile should be broken down into capital investment and ongoing operational costs. Consideration should be given to what funding sources are appropriate for each of these areas.

• A specific short-term plan should be developed for the capitalization of the new hardware required for regional networks.

• Efforts should be made to rapidly decrease recurring telemetry costs and to apply the savings to capitalization of modern telemetry equipment.

• A long-term plan should be developed for continued operation of regional networks. Special emphasis should be given to means of stimulating state funding for operational support for monitoring of regional seismicity.

• Strategies should be developed to take advantage of the improved data that will be available to generate new initiatives for support of research in regional seismology.

NATIONAL ACADEMY PRESS

The National Academy Press was
created by the National Academy of
Sciences to publish the reports issued by
the Academy and the National Academy
of Engineering, the Institute of Medicine, and the National Research Council, all operating under the charter
granted to the National Academy of
Sciences by the Congress of the United
States.

ISBN 0-309-04291-